SOVIET POWER: ENERGY RESOURCES, PRODUCTION AND POTENTIALS

SOVIET POWER:
ENERGY RESOURCES, PRODUCTION AND POTENTIALS

JORDAN A. HODGKINS
Kent State University

GREENWOOD PRESS, PUBLISHERS
WESTPORT, CONNECTICUT

Library of Congress Cataloging in Publication Data

Hodgkins, Jordan Atwood, 1920-
Soviet power.

Reprint of the ed. published by Prentice-Hall, Englewood Cliffs, N. J.
Includes bibliographical references.
1. Power resources--Russia. I. Title.
[HD9502.R82H63 1975] 333.7 75-31801
ISBN 0-8371-8491-6

Originally published in 1961 by Prentice-Hall, Inc., Englewood Cliffs, N.J.

Reprinted with the permission of Jordan A. Hodgkins

Reprinted in 1975 by Greenwood Press, a division of Williamhouse-Regency Inc.

Library of Congress Catalog Card Number 75-31801

ISBN 0-8371-8491-6

Printed in the United States of America

TO JANE and PRISCILLA

whose silence made this possible

PREFACE

The continued rapid rise in Soviet industrial output, and more especially, the impressive increases in the heavy industry sector, have aroused wide public interest in the non-Soviet world. In the United States, accustomed for so long to an unchallenged position as the world's leading industrial power, the rate at which the U.S.S.R. is cutting down the American lead has been a matter of no little concern. Those who, a decade ago, claimed to welcome Soviet competition in the economic arena are now not so sure.

Whether the Soviet Union will overtake the United States in overall industrial output in the early 1970's, as Mr. Khrushchev confidently claims, depends obviously not only on the continued effectiveness of the Soviet industrial machine, but also on the response of the U.S. economy to the challenge. On the latter there will be much argument.

But none will dispute that the ability of the Soviet Union to continue its present rate of industrial growth in the decades to come will depend to a very considerable degree on its mineral resources—their variety, their amounts, and their distribution. For the very nature of the Soviet state precludes it from becoming dependent on foreign sources of supply of essential minerals, as the United States, to an increasing extent, has come to be.

A decade ago Demitri Shimkin, at the end of an exhaustive survey of what was then known of Soviet mineral resources *(Minerals: A Key to Soviet Power)*, concluded that the mineral wealth of the Soviet Union, as then known or inferred, approximated that of the United States. Since then, U.S. resources have been consumed at an almost frightening rate by the ever increasing demands of industry, while new discoveries have not added reserves at a comparable rate. In the U.S.S.R., on the other hand, consumption of minerals, while it has increased rapidly also during the past decade, has in most cases been far less in amount than in the United States. At the same time, as the result of a greatly stepped-up program of geological exploration and prospecting, the U.S.S.R. has added greatly to the reserves of minerals it was already known to be rich in and has significantly bettered its position regarding minerals it was thought to be deficient in or lacking in altogether a decade ago. For example, industrial diamonds, the lack of which was painfully felt a few years ago in the Soviet machine-tool and metal-working industries, have been discovered in Eastern Siberia in quantities adequate to meet Soviet requirements of this strategic and valuable mineral.

Although there are still deficiencies in the Soviet mineral position, the over-all picture is one of great variety and, in the case of a number of important minerals, large reserves. In both respects, the U.S.S.R. is today clearly ahead of all other nations, including the United States. This is not surprising, considering the size of the country and its geological variety. At the same time, one cannot help but be impressed by the magnitude of the effort which has been concentrated on the geological exploration and prospecting which has revealed the riches previously only suspected. The importance attached by the regime to the inventorying of the country's mineral resources is reflected in the elevation of the former Geological and Prospecting Service to the all-Union Ministry of Geology and by its retention as one of the few all-Union ministries which were not decentralized among the Union republics and the new economic administrative *rayons* during Khrushchev's economic decentralization of 1957.

Fortunately for students of Soviet industrial capabilities, the great increase in geological exploration and prospecting has been accompanied by a substantial, if not equal, increase in the amount of published material on Soviet minerals by Soviet specialists. And while production figures for a number of minerals are still not published, for others they are reasonably detailed and up-to-date. Consequently, the foreign observer of the Soviet scene is able to obtain a considerably clearer picture of the Soviet mineral position than was possible a decade ago when the Stalin censorship was still in full sway.

When one compares the abundance of the Soviet Union's mineral wealth with the inadequacy of good agricultural land for its large and rapidly growing population, one cannot help wondering if the Soviet Union will not find it expedient, in the long run, to adopt the course which Britain embarked on over a century ago: export minerals and manufactured goods and import foodstuffs and agricultural raw materials—on an incomparably greater scale, of course. The system of totalitarian planning and absolute governmental control over natural resources and labor which has enabled the Soviet Union to develop its industry at such a rapid rate should also enable it to produce minerals and manufacture goods for export in such volume and to sell them at such price as would be necessary to capture a given foreign market. Thus, the pattern of natural resources would seem to provide a certain compulsion—or at least a kind of *rationale*—for a program of economic warfare which fits in admirably with the regime's professed political aims. There are indications already that this pattern of trade relationships is developing.

However, if we confine our view to the more immediate Soviet industrial goals, those of the current Seven Year Plan which terminates at the end of 1965, we find that the increase in electric power production which will be needed to achieve these goals is impressive enough: from an output of 233 billion kilowatt-hours in 1958 to between 500 and 520 billion in 1965. While a part of this increase will come from the spectacular new hydroelectric plants such as those—the world's largest—already completed on the Volga above Kuibyshev and at Stalingrad and the even more powerful plants under construction in Eastern Siberia at Bratsk on the Angara and Krasnoyarsk on the Yenisei, most of it—approximately four-fifths—will be provided by thermal electric plants which will require some kind of fuel—coal, oil, gas, peat, or oil-shale. Thus, the expansion of electric power production on which the continued rapid growth of industry is based, depends on a corresponding increase in fuel production—solid, liquid, and gas.

But Soviet industrial expansion depends on these minerals not only as sources of power, but also as sources of raw materials for the Soviet metallurgical and chemical industries. The production of coal for metallurgical coke, primarily for the iron and steel industry—on which the entire industrial structure rests—and which is scheduled to increase the production of pig iron and steel by approximately two-thirds, will have to reach 150-156 million metric tons in 1965 if the needs of ferrous and non-ferrous metallurgy are to be met. And the demands of the chemical industries, on the development of which the current plan lays especial emphasis, adds another dimension to the vital importance of coal and petroleum, namely, their importance as two of the most important raw materials for that branch of industry.

Clearly, a knowledge of the Soviet Union's resources and production potential of coal, petroleum, gas and the lesser solid fuels is essential for a critical evaluation of Soviet industrial prospects. A more liberal Soviet policy regarding the publication of data about its mineral resources has opened the door for the Russian-reading specialist to acquire this knowledge. In this book Dr. Hodgkins makes available the results of his painstaking researches in the already voluminous Soviet literature on mineral fuels. It should be especially welcome to all those concerned with Soviet industrial development who, because of the language problem, cannot consult the Soviet literature, or who do not have access to it.

John A. Morrison

Madison, Wisconsin

ACKNOWLEDGMENTS

The author desires to express his indebtedness to the many colleagues who have so substantially contributed to the completion of this book. I am especially grateful to Dr. Burton W. Adkinson, National Science Foundation, for his aid in securing ephemeral source materials; Dr. George B. Cressey, Maxwell Professor of Geography, Syracuse University, for his intellectual guidance and encouragement; Dr. John A. Morrison, Department of Geography, University of Pittsburgh, for his unfailing interest, scholarly evaluation, and contributions, including the authorship of the preface to this book; Dr. Erich Bordne and Dr. Hibberd V. B. Kline, Department of Geography, University of Pittsburgh, whose academic interests fostered the publication of this book; Dr. Earl C. Case, Professor Emeritus, University of Cincinnati, and Dr. Henry M. Kendall, Miami University, Oxford, Ohio, for their encouragement over the years; Dr. H. F. Raup and colleagues, Kent State University, for their many helpful suggestions; and Phyllis Wheeler Hodgkins, who has patiently aided and inspired this study.

Errors of fact or interpretation which may be contained herein are the sole responsibility of the writer, and can in no way be attributed to any of the above persons.

J. A. H.

CONTENTS

OIL SHALE APPENDIX

TABLES

FIGURES

INTRODUCTION

Energy resources and their development are the key to a nation's technical and industrial capacity and the resultant well-being of her peoples. In developing energy resources, man establishes geographic patterns which appear primarily in industrial and manufacturing regions. Changes in the further exploitation of these resources are reflected in the creation of new power bases. This is especially true in the Soviet Union where energy resources are currently undergoing extensive development.

One of the main goals of policy makers in the Soviet Union is to surpass the United States in industrial production. The attainment of this goal is directly dependent upon basic energy potentials. Since World War II, the Soviet government has undertaken an extensive program of economic expansion, budgeting as much in the current Seven Year Plan for this purpose as it did in the total preceding six Five Year Plans. In this study, the actual power base upon which this expansion must be built is examined in detail, region by region.

Specifically, the purpose of this report is a regional presentation of the potential and established base of nonrenewable energy resources in the U.S.S.R. These resources include coal and brown coal, oil shale, oil, and also natural gas. To fulfill this objective and facilitate comparisons between the various fuels, the potential energy base of each fuel is expressed in a single unit of energy measurement. Each energy resource is presented in terms of its kilowatt hour energy potential. In this way, the true relative position of each resource is readily discernible and not obscured by a comparison of tons of coal with barrels of oil or cubic meters of gas. It is also possible under this system to compare annual power production with potential and to present total regional power potentials of all fuel resources cartographically in a single unit of measurement.

To succeed in this objective and obtain a clear picture of Soviet power potentials, the following steps were necessary: (1) determine and evaluate the actual reserve base for each resource; (2) convert all reserve data on resources to their kilowatt hour content; (3) determine the established annual production of each resource; (4) compare the potential energy resource base with the established production base to ascertain the status of regional use, and to locate undeveloped potential power regions.

Covering as much of the earth's surface as North America, from the Panama Canal north through the Arctic, the vast territory of the U.S.S.R. is increasingly making its impact felt on the world. Yet its potentials and the ideologies by which they are formulated into practical land use realities are little understood. These potentialities, both human and natural resources, are tremendous and their combined ability to influence the future of our globe demands understanding. It is hoped that this study may in some measure contribute to that understanding.

The reserve base for each individual energy resource was compiled from Soviet sources. Specific sources are cited throughout the text. All data on the reserve base are presented in the metric system in accordance with original sources and to facilitate computations.

When the quantitative values, that is, metric tons or cubic meters, had been compiled, they were then converted to their calorie content. Conversions were computed for each separate basin and deposit, or groups of adjacent deposits, in order to obtain a regional pattern. Generalized thermal values were necessary as a step in computing the energy content as expressed in kilowatt hours. Thus, calorific values of all energy resources were computed. Soviet sources allocate the following calorie values to each of the energy resources: hard coal—7,000 kilocalories per kilogram; brown coal—2,500 kilocalories per kilogram; oil shale—1,900 kilocalories per kilogram; oil—10,000 kilocalories per kilogram; natural gas—8,800 kilocalories per kilogram; and peat—2,800 kilocalories per kilogram.[1] Because these values were based on analyses of Soviet energy raw materials in particular, they are the ones used in this study.

Similarly, once the thermal value for an energy resource had been determined, it was then converted to its kilowatt hour energy content. The standard conversion factor of ×.0011628 was utilized for

[1]Yu. Vasilev and K. Posplelova, "Structural Change in the Fuel Balance of the U.S.A.," *Planovoe Khozyaistvo*, No. 6 (1957), 68, and V. Kalamkarov, "Basic Trends in the Development of the Production of Fuel in the Sixth Five Year Plan," *Planovoe Khozyaistvo*, No. 4 (1957), 17.

this purpose.[2] This conversion factor is internationally accepted by agreement among most of the major nations of the world.

After conversions were completed, the kilowatt hour energy potential for each energy resource was plotted on a separate map of the Soviet Union. When all maps had been compiled, they were superimposed upon each other and the resultant composite data were evaluated and regions of varying energy potentials were determined. The composite map was compared with the individual maps on the energy production base as an aid in analyzing present and undeveloped power bases.

Data for this study, with the exception of bibliographic and comparative materials, were procured from sources published in the Soviet Union. These sources may be arranged in five categories: (1) geographic materials, principally regional monographs and journals; (2) reports on mineral deposits and reserves, including monographs, journals, serials and continuations; (3) data on economic planning and plans; (4) statistical compilations and reports; and (5) reports on current activities, including newspapers and industrial technical journals.

Uniformity of quality, reliability, and availability is not a characteristic of source material over the 40 odd years of Soviet power. Data can roughly be categorized by three time periods: (1) pre-1939; (2) 1940 to approximately 1954; and (3) 1955 to the present time. Until 1939, availability was excellent; libraries with an interest and funds were able to accumulate substantial collections in this country. Quality and reliability of the pre-1939 sources, however, are not of the same high standards maintained in the current Soviet scientific publications. Lack of precise definition of terms is one of the major irritations encountered in working with the material. This is a generalization, however, which should not be applied to the work of all Soviet scientists and technicians.

Reliable Soviet data is largely unavailable for the period 1940 through 1954. This is especially true of statistical reports and monographic works, as well as many journals. During this period data was presented in the form of percentage increases with undisclosed or vague base figures. Material can generally be described as boastful, but not revealing.

Since 1955 Soviet published materials have been readily obtainable. Availability should not be construed to mean that these published sources provide the student with all of the desired information or definitive data. On the contrary, the degree of detailed information varies greatly with the subject material. Quality analyses of the

[2] Norbert A. Lang, *Handbook of Chemistry* (Sandusky: Handbook Publishers, Inc., 1944), p. 1720.

chemical and physical properties of various raw energy materials are available in minute and overwhelming detail. Quantitative data on reserves of energy sources, other than coal and shale, are difficult to obtain for specific deposits, and must either be computed from scattered reports or in some instances, estimated. However, the quality of the work available since 1955 has been good to excellent in most cases.

According to Demitri Shimkin,[3] statistical reports on Soviet production are reliable to within less than 5 per cent error, but in using the material, careful scrutiny of definitions is necessary. For example, the English edition of "National Economy of the U.S.S.R." lists coal production as 391 million tons, but omits coal used in the production of coal, and coal mined by agencies other than the Ministry of Coal. This statistical handbook lists all coal produced. The regional distribution and difference can be ascertained by consulting statistical handbooks on the individual republics. This type of problem is frequently encountered in using Soviet statistical material.

[3]Demitri B. Shimkin, *Minerals: A Key to Soviet Power* (Cambridge: Harvard University Press, 1953), p. 10.

Chapter 1

SOVIET COAL: THE GEOGRAPHICAL DISTRIBUTION OF RESERVES, AND ENERGY POTENTIAL

The Soviet Union has 4×10^{12} tons more coal than was credited to the entire world in post World War II studies. Geological surveys computed for January 1, 1956, credit the Soviets with coal reserves totaling over 8.6×10^{12} tons (see Table 1).

Contrary to trends in the rest of the world, recent Soviet discoveries have resulted in tremendous increases in reserves. Estimates of total geological reserves for the world declined markedly in the decade after 1937. By 1948, this decline amounted to 2.3×10^{12} tons, a decrease from 7.9×10^{12} tons to 5.6×10^{12} tons. [1] Both the 1937 and 1948 estimates allocated 21 per cent of the world's coal to the Soviet Union. [2] Ironically, the same year that world coal reserves were revised downward, the Soviets undertook extensive surveys of their own coal resources. [3] As a result, the Soviets now claim to have 57 per cent of the total geological reserves of coal in the world. [4] The United States has 2.3×10^{12} tons of coal, or 6.3×10^{12} tons less than the Soviet Union.

Mineable reserves of Soviet coal are, reportedly, over 7.7×10^{12} tons, an amount sufficient to last 21 thousand years at the 1955 rate of mining or 12 thousand years at the projected 1970 production goal set by Mr. Khrushchev.

Mineable coal, or coal that the Soviets consider worth mining, is based upon the criteria of depth, thickness of coal seams, and its ash content. [5] These elements vary with the individual coal deposits. Certain limits, however, are prescribed for all basins. Hard coal must occur in seams from 0.45 to one meter in thickness to be considered mineable. Brown coal seams must be from 0.7 to one meter in thickness, with a specific minimum of 0.9 meter for the Moscow Basin. The maximum ash content for mineable coal ranges from 40-50 per cent depending upon depth and thickness of the coal seam,

TABLE 1

A COMPARISON OF GEOLOGICAL AND MINEABLE RESERVES OF SOVIET COAL: 1937 AND JAN. 1, 1956, IN TONS 10^9

Categories	Geological Reserves 1937[a]	Geological Reserves 1956[b]	Increase in Reserves 1937-1956[c]	Mineable Reserves 1956[d]
Total coal	1,654.00	8,699.51	7,045.51	7,765.29
Hard coal	1,443.00	5,656.60	4,213.60	4,903.53
Brown coal	221.00	3,012.91	2,801.91	2,861.76
Reliability				
Proven	131.30	241.21	109.91	223.05[e]
Probable	294.59	941.89	647.30	861.85
Possible	1,228.11	7,486.41	6,258.30	6,680.39
Depth Meters				
0 - 300	1,268.42	2,351.51	2,862.97	2,088.86
300 - 600		1,779.88		1,599.62
600 - 1,200	247.71	2,838.03	2,590.32	2,540.66
1,200 - 1,800	138.23	1,700.09	1,561.86	1,539.83

[a]M. M. Prigorovsky, "Coal Bearing Provinces and Basins of the U.S.S.R.," *Report of the XVII Session, International Geological Congress*, I (Moscow: 1939), p. 194.

[b]From Coal Appendix – Table I.

[c]Computed from columns 1 and 2.

[d]From Coal Appendix – Table II.

[e]N. V. Shabarova and A. V. Tyshnova, eds., *Zapasy Ugley i Goryuchikh Slantsev S.S.S.R.* (Moskva: Gosgeolteknizdat, 1958), p. 18.

with the exception of the Moscow Basin where the permissive limit has been set at 60 per cent.

Mineable reserves do not include the deeply deposited coal beds. In only two instances do they include deposits to a depth of 1,200 meters; these are the Donets and Kizel Basins. In all other hard coal basins, mineable reserves are computed only to the 600 meter depth. Most brown coal basins do not have mineable reserves below the 300 meter level.

Geographic location, transportation facilities, and hydrogeological conditions were not considered by Soviet geologists in making these computations. Thus, mineable coal constitutes what is often called *the balance of reserves* in the Soviet Union. It should be noted that the term *mineable* refers only to the coal which the Soviets believe to be worth mining in a basin. It is not an estimate of what can be extracted.

When measured in terms of its energy potential, the mineable coal of Soviet Russia contains 48×10^{15} kilowatt hours. Bituminous and anthracite, in all their forms, constitute 83 per cent of the total energy potential, and brown coal but 17 per cent. If coal alone were used to produce electricity, this potential would be equivalent to a 282 thousand year supply, at the 1955 rate at which all energy resources produced electricity.

Vast differences in energy potentials exist between coal basins having approximately the same reserve tonnages of coal but different types of coal. Yet when considered within the same regional framework as the distribution of geological reserves, the energy potentials evidence the same spatial pattern. Many negating factors, such as the ash, moisture and sulfur content of the coal, its depth, its geographical and geological accessibility, determine the effectiveness of this pattern.

Since 1913, coal production has risen 1,597 per cent, from 29.1 $\times 10^6$ tons to 506.5 $\times 10^6$ tons in 1959, or from 440 kilograms per person to 2.4 tons per person. This is a result of three important production trends. Firstly, there has been the migration of mining activity from European Russia to the eastern regions. Secondly, there has been a rise in mining by the open pit method; in 1959 the Soviets mined 20.2 per cent of all coal by the open pit method, compared with 0.6 per cent in 1913. Thirdly, brown coal production and its utilization as a local fuel has increased in all regions; in 1913 only 3.9 per cent of the coal produced was brown coal; in 1959, 27.9 per cent was brown coal.

Classification of Soviet Coal.—Terminology for the classification and quality of coal varies from country to country and within countries. [6] Soviet classification for metamorphosed (hard) coals varies with the individual basins. However, most Soviet coal can be equated to the classification system applied to Donets and Kuznetsk grades of coal as presented below. [7]

Name of Grade	Symbol	Volatiles %	Carbon %	Assay Characteristics
Dry long flame	D	Above 42	75-86	noncaking, powdery to conglomerate
Gassy	G	36-44	78-89	caking, fusiable, occasionally crumbly
Fatty steam	PZH	26-36	84-90	caking, fusiable, firm, moderately thick
Coking coal	K	18-26	87-92	caking, fusiable, compact, moderately compact
Dry steam	PS	12-18	88-94	caking or fused, partly compact
Lean	T	below 17	90-95	noncaking, powdery or conglomerate
Anthracite	A	below 8	----	--------------

While this classification can be used with most Soviet coals, it is inconsistent with our own coal terminology. The Soviet term *hard coal* designates all coals other than the brown coals; *brown coal* includes our classification of lignites and subbituminous coals [8] (see Coal Appendix - Table III).

Coal Reserves of the U.S.S.R. on January 1, 1956.—Geological reserves of Soviet coal are 8.69×10^{12} tons. This represents an increase of 7×10^{12} tons, or 426 per cent over the previously computed and announced reserves (see Table 2). Sixty-five per cent are classified as hard coal, and 35 per cent as brown coal. In 1937, brown coal constituted 13 per cent of the reserves and hard coal 87 per cent.

Approximately 3 per cent of the total Soviet coal reserves are valid, or proven reserves; 11 per cent are presently classified as probable; and 86 per cent are listed as possible, or inferred reserves. Since 1937, reserves in the proven category have increased by 85 per cent, and inferred reserves by 429 per cent. Most of the surveying has been of the reconnaissance type.

A total of 2.3×10^{12} tons or 27 per cent of the Soviet coal reserves are within 300 meters of the earth's surface. This sum is equivalent to the total reserves for the United States. Twenty per cent of the reserves are located between 300 and 600 meters below the surface of the earth; and 53 per cent occur at depths of 600 to 1,200 meters. Another 20 per cent exist between 1,200 to 1,800 meters below the earth's surface, but it is only in the Donets and Kizel Basins that coal is considered mineable at these depths.

Distribution and Quality of Geological and Mineable Reserves.—Soviet Russia's enormous coal reserves are dispersed throughout 173 basins and deposits (see Figure 1). In size they range from a single basin, the Lena, which reputedly has more coal than all of North America, to scattered deposits of a few million tons each.

Among the traditionally established regions of the Soviet Union, the geological reserves are distributed as follows:

Region	Per Cent of Geological Reserves	Region	Per Cent of Geological Reserves
European	7.50	Arctic, Subarctic Siberia	55.50
Urals	0.08		
Caucasus	0.02	South Siberian belt	31.40
Kazakhstan	1.60		
Middle Asia	0.40	Far East	0.60
Transbaikal	0.10	North East	2.80

Over 60 per cent of these reserves are in basins which lack established industrial complexes, and have no railroad connections with the rest of the country. Seven of the 173 basins and deposits contain 88.6 per cent of the total reserves (see Fig. 2). Two of these seven basins, the Lena and the Tunguska, have half of the total reserves. The famed Donets and Kuznetsk basins, as listed below, collectively contain but 13.3 per cent of the reserves.

Basins	Tons (10^9)	Per Cent of Total Geological Reserves
Lena	2,647.	30.5
Tunguska	1,745.	20.1
Kansk-Achinsk	1,220.	14.1
Kuznetsk	905.	10.5
Taimyr	583.	6.7
Pechora	344.	4.0
Donets	240.	2.8
Total	7,684.	88.7

Coals of the Lena, Tunguska, and Taimyr basins, regardless of their quality, will be restricted to local use in the foreseeable future.

Significant regional differences are evident in the quality of Soviet coals. These differences determine the place and manner in which coal will be used. Undesirable qualities of coal are a high moisture, ash, and sulfur content. Desirable qualities are a high carbon and calorie content. Volatile gases when present are useful byproducts. According to Soviet standards, an ash content above 40-50 per cent places the material in the combustible shale category. [9] Calorie content, or heating values, referred to in this section of the text is for the combustible masses only (see Coal Appendix - Table III).

European Russia.—European Russia with fourteen basins and deposits has 647.28×10^9 tons of coal, or seven and one half per cent of the total geological reserves. [10] Present figures represent substantial increases in the reserves of known basins, and also include two new basins, the Lvov-Volyn and the Kama.

Reserves of coal in the Donets Basin are now reported as 240.62×10^9 tons, a 173 per cent increase above previous estimates. [11] Mineable coal, according to Soviet computations, is approximately 190×10^9 tons. Donets coals are of high quality; they have a high calorie and carbon content, and low moisture and volatile content (see Coal Appendix - Table III, Characteristics of Soviet Coals). Shevchenko sums up their undesirable qualities by stating that: "Unfavorable characteristics of Donets coals are their high ash content and large sulfur content". [12] The ash content fluctuates widely from 2 to 32 per cent dry coal. However, it responds well to benefication and the

COAL BASINS AND DEPOSITS OF THE U.S.S.R. – LOCATIONAL KEY TO FIGURE 1

1. Lvov-Volyn Coal Region
2. Deposits of the Trans-Carpathian Coal Area
3. Carpathian Coal Area
4. Deposits of the Dnestrov Coal Area
5. Bolgrad Deposits
6. Dneper Basin
7. Deposits of the Crimea
8. Moscow Basin
9. Donets Basin
10. North Caucasus Deposits (Carboniferous)
11. North Caucasus Deposits (Jurasic)
12. Tkvarcheli Deposit
13. Tkibuli Deposit
14. Akhaltsikh Deposit
15. Timan Group of Deposits
16. Pechora Basin
17. Paikhoya Deposits
18. Deposits of Shchugor-Vuktyl Region
19. Kizel Basin
20. Kama Basin
21. Ufimsk Group of Deposits
22. South Ural Basin
23. North-Sos Vinsk Coal Region
24. Serov Coal Region
25. Bulansh-Elkinsk Deposits
26. Chelyabinsk Basin
27. Deposits of the East Slope of the Urals (Egorshino hard Coal Region, Makhnev Deposit, Poltavo-Bredin Coal Region)
28. Orsk Basin
29. Ural-Caspian Basin
30. Inder Group of Deposits
31. Deposits of the Ural-Emba Region
32. Berchogursk Deposit
33. Mangyshlaka Deposits
34. Tuarkyrsk Coal
35. Balkhan Coal Region
36. Kugi-Tang Deposit
37. Baisun Deposit
38. Kshtut-Zauran Deposit
39. Fan-Yagnob Deposit
40. Shargun Deposit
41. Shuroabad Deposit
42. Ravnous Deposit
43. Mianadus Deposit
44. Nazar-Ailoksk Deposit
45. Sulyukta Deposit
46. Shurab Deposit
47. Kyzyl Kiya Deposit
48. Aldyyar Deposit
49. East Fergana (Uzgen) Basin
50. Kok-Yangak Deposit
51. Narynsk Deposit
52. Angren Deposit
53. Lenger Deposit
54. Kara-Tau Coal Area
55. Kok-Mainak-Karakichin Coal Area
56. Sogutin Deposit
57. Dzhergalan Deposit
58. Turgay (Ubagan) Basin
59. Dzhezkazgan Coal Region

60-61-62. Deposits of the Teniz-Kipchak Coal Region

63. Deposits of the Akmolinsk-Kokchetav Coal Region
64. Zavyalov Deposit
65. Karaganda Basin
66. Kuu-Chekin Deposit

67-68. Yeremen Tau Coal Region

69. Ekibastuz Deposit
70. Maikyuben Deposit
71. Bayan-Aul Coal Region

72-73. Dzhamanguz Coal Region

74. Bala-Saran Deposit
75. Irtysh Deposits
76. Western Balkhash Coal Region
77. Kemelbek and Ushkuduk Deposits
78. Alakul Deposit
79. Kara-Say-Mukrin Deposit
80. Oy-Karagoy Deposit
81. Ayaguz Group of Deposits
82. Tarbagatay Group of Deposits
83. Zaisan Group of Deposits
84. Mulnay Group of Deposits
85. Kosh-Agach Group of Deposits
86. Altay Group of Deposits
87. Kuznetsk Basin
88. Gorlov Basin and Ob Deposits
89. Tomsk Group of Deposits
90. Kansk-Achinsk Basin
91. Minusinsk Basin
92. Ulukhem Basin
93. Intital Deposit
94. Aktal Deposit
95. Irkutsk Basin
96. Khakharey Deposits
97. Deposits of the Irkut River Valley
98. Tunguska Basin
99. Ust-Yenisey Coal Region
100. Taimyr Basin
101. North Taimyr Basin
102. Baikal Group of Deposits
103. Dzhidin Group of Deposits
104. Lake Gusin Deposit
105. Deposits of the Udin Valley
106. Tugnuy Deposit
107. Deposits of the Khilok Region
108. Tarbagatay Deposit
109. Deposits of the Padin Depression
110. Urey Deposit
111. Alenguy Deposit
112. Chernov Deposit
113. Lake Sosnov Deposit
114. Kharanor Deposit
115. Matsiev Group of Deposits
116. Duroev Deposit
117. Arbagaro-Kholbon Deposits
118. Delyun Deposit
119. Staro-Olov Deposit
120. Bukachachan Deposit
121. South Yakutsk Basin
122. Nyukzhin Group of Deposits
123. Tolbuzin Deposit
124. Verkhne-Zey Group of Deposits
125. Depsk and Novo-Yampol Deposits
126. Kivda-Raichikhinsk Coal Region
127. Ushumun Deposit
128. Tyrmin Coal Deposit
129. Bureya Coal Basin
130. Ogodzhinsk Deposits
131. Deposits of the Urmiy Coal Region
132. Khabarovsk Coal Area
133. Khungar Deposits
134. Amgun-Amur Group of Deposits
135. Sovgavan Group of Deposits
136. Bikinsk Deposit
137. Khankay Coal Region
138. Suifun Basin
139. Uglov Basin
140. Suchan Basin
141. Daubikhinsk Coal Region
142. Uglenosnaya Area of the Upper Iman Region
143. Deposits of South Sakhalin
144. Deposits of the Poronay Coal Region
145. Deposits of the Uglegorsk Region
146. Deposits of the Krasnogorsk Region
147. Deposits of the Aleksandrovsk Region
148. North West Coal Region
149. East Sakhalin Coal Region
150. Deposits of the Shmidt Peninsula
151. Lena Basin
152. Zyryansk Coal Region
153. Dzhelkan Deposit
154. Arkagalin Coal Area
155. Darpirsk Deposit
156. Chelemdzhin Deposit
157. Pervomaisk Deposit
158. Elgen Deposit
159. Omsukchan Coal Area
160. Okhotsk Coal Area
161. Avekov (Gizhiginsk) Deposit
162. Omolon Coal Area
163. Deposits of the Penzhinsk Region
164. Tigel Deposits
165. Podkagernoe Deposits
166. Krutogorov Deposit
167. Deposits of the Kharyuzov Region
168. Nyapan Deposit
169. Chaun-Chukotsk Coal Area
170. Anadyr Coal Area
171. Deposits of the Bukhty Coal Region
172. Tiksi Coal Deposit
173. Deposits of Franz Joseph Land

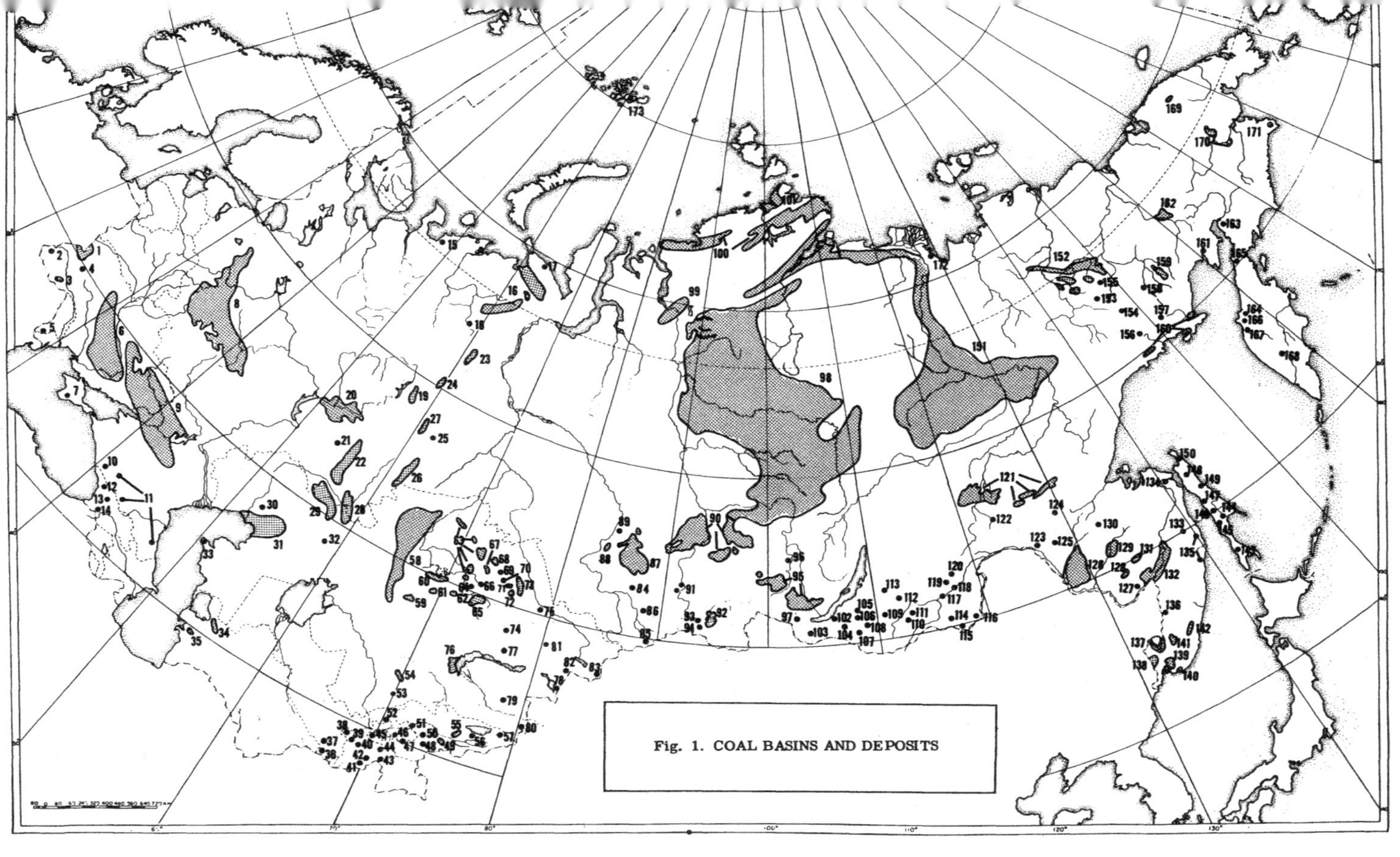

Fig. 1. COAL BASINS AND DEPOSITS

ash content of coal shipped to other areas for consumption is low compared to that for other basins (see Table 3).

Sulfur in Donets coal ranges from 0.4 to 7.3 per cent. Four categories of sulfur content in Donets coal include: low sulfur—below 1.5 per cent; average sulfur—1.5 to 2.5 per cent; considerable sulfur—2.5 to 4.0 per cent; and high sulfur—more than 4.0 per cent. Only 10.3 per cent of the coal in the Donets basin falls within the low sulfur classification. Pyrites are the predominant form of sulfur in the coal and the larger ones may be removed during concentration.

The largest basin in European Russia is the Pechora with 344.5×10^9 tons of hard coal, of which 262×10^9 tons are considered mineable. This represents an 855 per cent increase in geological reserves, the greatest increase in the European section. In general, Pechora coals have a slightly higher calorie and carbon content than Donets coals, but volatile material for comparable types of coal is higher. The moisture content of air-dried Pechora coal is greater than that of the Donets. Absolutely dry coal has an ash content which ranges from 8.0 to 50.0 per cent, and in one deposit it exceeds the permissive limits and reaches 60 per cent. Coal shipped to other regions for consumption exceeds 19 per cent in ash content. Pechora coals have a sulfur content of 0.4 to 10.0 per cent. High sulfur coals are found in the Usinsk deposit only, and most coals of the Pechora basin do not have as high a sulfur content as that attributed to the Donets Basin.

Brown coal reserves of the Dneper Basin have increased by 706 per cent. Of the 4×10^9 tons credited to this basin, 3.66×10^9 are presumed mineable. Dneper brown coal has the highest moisture content in the Soviet Union. [13] Its ash content varies between 12 and 40.0 per cent for dry coal. The calorie and carbon contents are low while the sulfur content is average to high. Open pit mining and bricketing make its utilization feasible as a local fuel.

Moscow Basin brown coal reserves total 24×10^9 tons, 98 per cent more than the 1937 estimates. Slightly more than 17×10^9 tons are listed as mineable, but much of this may be reserved for underground gasification. [14] Moscow coal is characterized by a high moisture, ash, and sulfur content. [15] Its moisture content is lower than Dneper coal, but the ash content for dry coal is higher. In computing the reserves of the Moscow Basin only, an ash content of 60 per cent was used to distinguish coal from combustible shale. [16] Sulfurous Moscow coals range between 1 and 5 per cent. Carbon and calorie contents are higher than Dneper brown coal.

The Lvov-Volyn hard coal basin, adjacent to the Polish border in the Ukraine, has geological reserves of 1.78×10^9 tons; 1.42×10^9 are mineable. Considered industrially important because of its proximity to the coal-poor Baltic region, the development of this area has re-

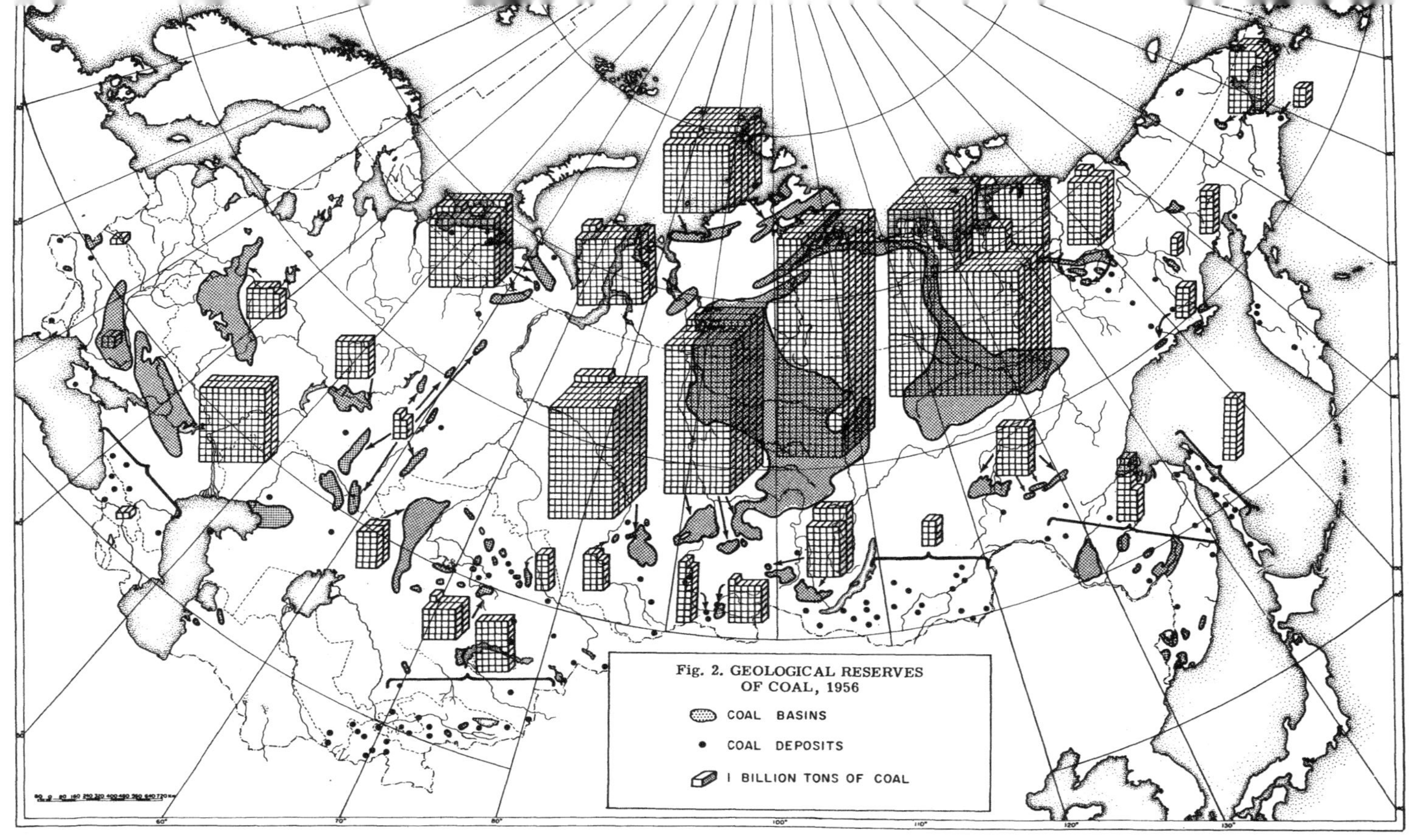

Fig. 2. GEOLOGICAL RESERVES OF COAL, 1956

ceived the personal attention of Mr. Khrushchev. [17] It has been mined since 1955. [18] Lvov-Volyn hard coal is lower in sulfur and ash than similar types of Donets coal, but higher in moisture content. Its heating capacity and carbon content are less than Donets coals.

Located near the confluence of the Kama-Volga Rivers, the Kama Basin contains over 30×10^9 tons of hard coal. Exploratory oilwell drilling revealed this new basin in 1952. [19] Although still in the process of being surveyed, geologists claim that nearly all the currently known reserves are mineable. Kama coal has a low calorie content, equal in heating capacity to the poorest of Donets coal. Moisture and ash contents are comparable to Donets coals. Sulfur content of the absolutely dry coal is 3.2 to 5.3 per cent.

Small and as yet undeveloped deposits of brown coal exist in the Moldavian Carpathians and the Carpathian foothills, [20] as well as in the Crimea. All Carpathian deposits have a total of 1.6×10^9 tons of coal. With the exception of the Transcarpathian deposits, there is no evidence of exploitation.

According to post World War II estimates, Europe's entire coal reserves, exclusive of the Soviet Union, total 648.81×10^9 tons, an amount approximately equal to those of European Russia.

The Urals.—Ural coal reserves, distributed among ten industrial basins and deposits, comprise only 0.08 per cent of the total Soviet geological reserves. Brown coal predominates, constituting 78.6 per cent of the 7.51×10^9 tons. The 1956 estimates of geological reserves are 13×10^6 tons less than the 1937 estimates for the Urals.

Kizel on the Western Slope of the Urals possesses 66 per cent of the 1.6×10^9 tons of hard coal reserves. Nearly all of Kizel's 1.06×10^9 tons are mineable; the coal has a high volatile content and fairly high calorie content. It differs from coals of other basins in its high sulfur and ash content. The sulfur content of absolutely dry coal is 4.0 to 8.0 per cent with an average content of 6.4 per cent. [21] Dry coal has an ash content of 18.0 to 40.9 per cent. Even after concentration, the sulfur content is 4 per cent and above; and the ash content of coal shipped to consumers is 24.7 per cent.

The remaining hard coal reserves are located on the Eastern Slope of the Urals at Egorshino, Elkinsk, Makhnev, and Poltavo-Bredin; and in the Southern Urals in the vicinity of Dombarovski. Of the 549×10^6 tons of hard coal in these five deposits, none has a calorie content, but all are high in volatiles and ash. Dombarovski anthracite is the only deposit other than Kizel with a high sulfur content.

Brown coals are located in the South Ural Basin (1.76×10^9 tons), the Chelyabinsk Basin (1.63×10^9 tons), the Orsk-East Ural Basin (1.03×10^9 tons), and the North Sosvinsk coal region (1.07×10^9 tons). Surveys are in progress on the Ural Caspian Basin.

With the exception of the South Ural and the Ural-Caspian Basin, brown coals of the Urals have a lower sulfur and ash content than Kizel hard coal. Volatile and moisture contents are lower than Dneper or Moscow brown coals, and their heating capacity is higher. South Ural and Ural-Caspian brown coals have a high moisture, ash, and sulfur content. Since sulfur in the Ural-Caspian Basin varies from 7.2 to 9.0 per cent, this may explain Soviet reluctance in developing the basin.

Caucasus Mountains.—The seven coal areas of the Caucasus region comprise only 0.02 per cent of the total reserves. Geological reserves of these deposits total 2.01×10^9 tons, and 1.01×10^9 are mineable. Four of the seven deposits are on the North Slope of the Caucasus which has 1.32×10^9 tons of the geological reserves. Estimates of the present reserves are about half those of 1937.

Eighty-six per cent of the reserves are hard coal, including all four of the North Slope Deposits. [22] Two of the Georgian deposits, Tkibuli with 460×10^6 tons and Tkvarcheli with 80×10^6 tons are hard coal. Akhaltsikh, in Georgia, contains reserves of 150×10^6 tons of brown coal.

Analytical data on Caucasus coals are incomplete. Available data indicate that North Slope deposits do not have a good heating capacity and are low in sulfur content, but high in ash content. Georgian hard coals have a low sulfur content, and are high in ash. Their heating capacity is not great for hard coals. Coal shipped to consumers from the Caucasus has the highest ash content of any hard coal area in the western section of the Soviet Union. Akhaltsikh brown coal does not provide a great deal of heat per kilogram of combustible material. It is low in volatiles, has a considerable amount of sulfur, and is high in ash.

Kazakhstan.—Kazakhstan possesses 1.6 per cent of the Soviet Union's coal reserves. The 139.9×10^9 tons of coal in this region are located in thirty-one basins and deposits. [23] Fifty-five per cent of the reserves are hard coal and 45 per cent brown coal. Mineable reserves total 123×10^9 tons. Geological reserves have increased 138 per cent since 1937; however, much of the increase is due to the discovery of new brown coal basins.

Karaganda, the foremost basin of Kazakhstan, has reserves of 51.23×10^9 tons, or 53 per cent of the Republic's reserves. Brown coal constitutes 1.17×10^9 tons. Continuous surveys since 1939 have resulted in a downward revision of 1.46×10^9 tons in Karaganda's geological reserves. [24] Karaganda hard coals have as high a heating value and carbon content as similar grades of Donets coal. Volatile gases are less. Sulfur ranges from 0.4 to 3.5; it averages 1 per cent, which is much lower than Donets coal, but twice as high as Kuznetsk coal. The ash content of absolutely dry coal is 10.0 to 43.0

per cent. Since 1940, the ash content of coal shipped to consumers has risen from 18.5 per cent to 22.1 per cent. This is higher than Donets or Pechora coals, and more than twice that of Kuznetsk coal.

Ekibastuz, the second major hard coal basin in Kazakhstan, contains 12.21×10^9 tons, of which 10.76×10^9 are mineable. This basin has 144×10^6 tons of coal per square kilometer, which is the greatest density for any coal accumulations in the world. [25] The ash content ranges from 14.0 per cent to 35.0 per cent. Moisture in air dried coal varies between 5 and 12 per cent. The sulfur, volatile, and carbon contents, as well as the heating capacity are lower than similar Karaganda coal.

Ubagan, located in the Turgay Depression of Western Kazakhstan, is the most recently discovered basin in this region. Reserves of brown coal total 36.49×10^9 tons, and are higher in quality than the Chelyabinsk brown coals to the west. [26] Mineable reserves, most of which are within 300 meters of the surface, total 35.4×10^9 tons. Sulfur in absolutely dry coal of the Ubagan Basin attains a maximum of 2 per cent which is half the minimum content of Kizel coal, and much less than that found in Dneper or Moscow coals. The moisture content is considerably less than the two European basins, but the ash content appears to be similar. Heating capacity is greater.

Maikyuben, with reserves of 21×10^9 tons, is the second largest brown coal deposit in Kazakhstan. Thirteen $\times 10^9$ tons are considered mineable. Calorie and carbon values are high, and the sulfur content does not exceed 1 per cent. The ash content exceeds most of the brown coals of the Urals but is less than either the Dneper or Moscow brown coals.

Lenger brown coal totals 2.02×10^9 tons. This deposit is located along the southern border of the Republic, northeast of Tashkent. Mineable coal is computed at 1.98×10^9 tons. The sulfur content ranges from 2.5 to 5.0 per cent; its ash content is comparable to the better brown coals of the Urals, and half that of Moscow brown coals. Volatiles are less than those found in Dneper, Moscow, or Ural brown coal deposits; heating capacity appears to be higher.

The balance of coal in Kazakhstan, 14.47×10^9 tons of hard coal, and 2.47×10^9 tons of brown coal, is scattered throughout twenty-six deposits. Eleven are nucleated around the Karaganda Basin, but the remainder are dispersed from the Chinese border in the east to the Caspian Sea in the west.

Middle Asia.—Middle Asia has 0.40 per cent of the total Soviet coal reserves. The 40.78×10^9 tons accredited to this region in 1956 represent an increase in reserves of 150 per cent over the 1937 figures. [27] Hard coal comprises 26.60×10^9 tons and brown coal 14.18×10^9 tons. Thirty-eight $\times 10^9$ tons are mineable. Numerous small scattered deposits characterize the coal accumulations of the re-

gion. Shabarova and Tyshnova state that there are 140 deposits of coal in Middle Asia, but list only 22 on their map. [28] Gapeev depicts 41 on his map but does not give reserve data. [29]
Major hard coal deposits are as follows:

Deposits	Geological Reserves (tons 10^9)	Mineable Reserves (tons 10^9)
Hissar coal region	3.78	1.96
South Hissar	1.67	1.67
Zeravshan	1.93	1.93
South Tadzhik	2.46	2.46
Ziddin	1.44	1.44
Magian	1.07	1.06
Fan-Yagnob	1.78	1.77
North Fergana	2.55	2.54
Kok-Yangak	2.07	2.07
Uzgen (East Fergana)	3.09	3.01

The quality of Middle Asian hard coals is excellent. Sulfur contents are low to average, with only two deposits reaching the considerable sulfur stage. Existing data indicate that air-dried coals are lower in moisture content than coals at either the Donets or Pechora Basins. Absolutely dry coal has an ash content comparable to Donets coals. In recent years the per cent of ash in coal shipped to consumers has declined from 19.8 per cent in 1945, to 14.9 per cent in 1955. Volatiles are low and the carbon content is high. Heating capacities are as good as Pechora or Donets coals.
Significant brown coal deposits exist at:

Deposits	Geological Reserves (tons 10^9)	Mineable Reserves (tons 10^9)
Angren	2.82	2.80
Shurab	3.08	2.82
Kyzyl-Kiya	2.38	2.27
Minkush	4.21	4.21

Although complete analytical data are lacking on the quality of Angren coal, it can be classified as the richest of the brown coals of Middle Asia. Shurab brown coal has the lowest sulfur content but the highest ash content. In general, brown coal of Middle Asia has a lower ash content than hard coal. Other than Minkush which has considerable sulfur, these coals may be considered as having an average sulfur content.

Siberia.—Western and Eastern Siberia can be called the power house of the Soviet Union because they contain 86.5 per cent of the country's geological reserves. Of the 7.5×10^{12} tons of coal found here, 4.8×10^{12} tons are hard coal and 2.7×10^{12} tons are brown coal. Arctic and Subarctic Siberia, or the region north of 58° North Latitude, have approximately 60 per cent of the total Soviet reserves; and the South Siberian Belt, from the fifty-eighth parallel south to the border and west from Lake Baikal to Novosibirsk, contains 26 per cent.

Arctic and Subarctic Siberia.—Arctic and Subarctic coal basins are noted for their isolation from the rest of the country. There are no known rail connections from all seven of the basins in this region.

The Lena, with its 2.6×10^{12} tons of coal is the largest basin in the Siberian Arctic and the Union. Although most of this coal (2.5×10^{12} tons) is still in the inferred category of reserves, sufficient geological work has been done over the last decade to lend credibility to Soviet claims. [30] Current reserves exhibit a 1,204 per cent increase over the 1937 estimates. Variety of grades typifies the coal of this basin. Brown coal predominates, however, totaling over 1.5×10^{12} tons. Reportedly, 2×10^{12} tons of the total reserves are mineable.

Qualitative data, available for eight specific deposits of brown coal in the Lena Basin, indicate that the vast majority of this coal has a sulfur content below 1 per cent. Ash in absolutely dry coal is one half to one third less than Dneper or Moscow coals, and the maximum moisture content in dry coal is one third that of the European brown coals. Volatiles are less and the heating capacity appears higher. Lena Basin hard coals have a moisture content below 5 per cent and a sulfur content below 0.6 per cent, which is far superior to Donets, Pechora, or Kizel coals. Data are nonexistent on carbon content, but the calorie value indicates that it cannot be high.

Second largest of the Siberian Arctic basins is the Tunguska, with 1.7×10^{12} tons of hard coal. [31] Mineable reserves are 1.5×10^{12} tons. Recent survey data indicate an increase of 297 per cent in geological reserves above the 1937 figure. Tunguska hard coals are low in sulfur and moisture but tend to be high in ash content. When compared to similar coals of the Lvov-Volyn, Donets, and Pechora basins, they have a lower carbon and calorie value and higher volatile content.

Taimyr, the northernmost of the Siberian basins, has reserves of 583.5×10^9 tons, of which 557×10^9 tons are hard coal. More than 511×10^9 tons in this recently discovered basin are listed as mineable. Taimyr hard coals have a lower sulfur, ash, and moisture content than Donets, Pechora, or Karaganda coals. The carbon and calorie content is as good as Karaganda or Pechora coals, but less than Donets coals.

Located at the mouth of the Yenisey River is the 222×10^9 ton, Ust-Yenisey hard coal Basin. With 214×10^9 tons of mineable coal, it is the fourth largest North Siberian basin. Sulfur in Ust-Yenisey coal is less than one half of one per cent, and the ash content is less than 7 per cent. Volatiles, carbon and calorie contents are less than similar Donets or Pechora coals.

South Yakutsk, with its 40×10^9 tons of hard coal, is potentially the most important basin of the Subarctic region. Directives of the 20th Congress of the Communist Party considered this basin as being the most "...economically expedient for the creation of a ferrous metallurgical base in the East." [32] In addition to its 39×10^9 tons of mineable coal, the region contains a known 1.6×10^9 tons of iron ore. South Yakutsk coal is lower in moisture, ash, and sulfur than comparable Donets, Karaganda, or Pechora coals. Complete data are unavailable on the carbon content, but the heating capacity is high.

The brown coal deposits account for the remaining 56×10^6 tons of coal in the region. Both are on the Arctic coast, one on the Taimyr Peninsula, and the other near the Arctic port of Tiksi. Tiksi brown coal is low in sulfur, ash, and moisture and has a heating capacity equal to the best brown coals of the Urals.

The South Siberian Belt.—The South Siberian Belt, including Tuva, has fourteen basins and deposits with a total of 2.3×10^{12} tons of coal, or 26 per cent of the geological reserves of the Soviet Union. Fifty-three per cent of these reserves are brown coal.

Recent surveys determined that Kuznetsk, the most prominent basin in the region, has shown a 100 per cent increase in geological reserves. The most intensively studied basin in the Soviet Union, a special serial publication is devoted to its problems and the results of continuing survey work. [33] Reserves are now listed as 905×10^9 tons of hard coal; 804×10^9 tons are mineable. Coking coal constitutes 260×10^9 tons of geological reserves. [34] Kuznetsk hard coals are the highest quality coals in the Soviet Union. The sulfur content is low, averaging 0.5 per cent. [35] Ash content of dry coals is approximately half those of the Donets and more than a third less than those of the Pechora or Karaganda Basins. Coal shipped from the Kuznetsk Basin has an ash content of only 10.5 per cent. Carbon, volatiles, and heating capacities resemble Donets coals.

Geological reserves of the Irkutsk Basin are 88.9×10^9 tons, an increase of 7.6×10^9 tons since 1937. Hard coals dominate with 84.7×10^9 tons. Reportedly, 67.4×10^9 tons of hard and brown coal are classified as mineable. Phosphorus in one layer reaches 1.2 per cent. [36] Irkutsk hard coals are exceedingly high in sulfur, ranging from 0.5 to 8.0 per cent, which is 2 per cent higher than similar Donets coals. Experimental coking, using only the best coal concentrated by hand, still results in unsatisfactory coke with a high ash content from 10.1 to 15.5 per cent. [37] The heating capacity of the combustible mass is equal to comparable grades of coal in the Donets, Kuznetsk, or Pechora Basins.

Brown coal of the Irkutsk Basin has relatively low sulfur content, the maximum being 2 per cent. Its ash content, however, exceeds even the Dneper Basin maximum and is greater than any brown coal deposit in the Urals. Volatiles are lower than those of the European basins.

Minusinsk, the third largest hard coal basin in the region, has 36.9×10^9 tons of coal. Surveys conducted since 1937 have resulted in a 79 per cent increase in reserves. Practically all is considered mineable. Minusinsk coals are of the same types as the Irkutsk Basin but of higher quality. Sulfur and moisture contents are much lower, while ash and volatile contents are about the same. The carbon content is unknown; the heating capacity is greater than Irkutsk coal but not as high as Donets coal for the same types.

Gorlov, the smallest hard coal basin in the area, has had a 1,046 per cent increase in reserves. Present reserves figures are 17.2×10^9 tons, all of which is anthracite coal. Fifteen $\times 10^9$ tons are mineable. Anthracite coal of the Gorlov Basin is equal in quality to that of the Kuzbas.

Deposits of hard coal in the Tuvinian Autonomous Oblast total 18.7×10^9 tons. The Ulukhem Basin contains 10×10^9 tons of these reserves. Although it was discovered in 1888, systematic surveys of the basin were not instituted until after the territory was incorporated into the U.S.S.R. Surveying began in 1947 and was completed in 1955. Ulukhem coals have less moisture, sulfur, and ash than comparable Donets coals, but are slightly higher than Kuznetsk coals. Heating capacities are higher than the combustible mass of Donets or Kuznetsk coals. Carbon contents are unknown.

Two small deposits, the Intital and Aktal, contain the balance of reserves. Approximately 11×10^9 tons are mineable.

Forty per cent of the Soviet's brown coal is in the Kansk-Achinsk Basin. Geological reserves of 1.2×10^{12} tons place this basin third in size among Soviet coal basins. Its 1,335 per cent increase in reserves is the greatest increase of any basin in the Soviet Union. According to L. Semenov, the heating capacity of Kansk-Achinsk brown coal "ranks it first place among the brown coals mined in the U.S.S.R." [38] In addition to its high calorific value, it has a sulfur content which is less than 1 per cent, and an ash content not in excess of 20 per cent. Air-dried coal has a maximum moisture content of 25 per cent. This is removed by artificially drying the coal at temperatures of 250-300 degrees centigrade. During the drying process the coal becomes powdery and can be burned only in specially adapted furnaces in thermal electric stations.

The Kansk-Achinsk Basin has about 1.76×10^9 tons of hard coal in the Sayano-Partizan deposit but qualitative data are unavailable.

Brown coal, 2.9×10^9 tons, is also found in the Tomsk Oblast, and in several small deposits in the mountains of the Kuzbas. Collectively, the geological reserves of the South Kuzbas deposits total 11×10^6 tons.

The Transbaikal Region.—Seven major deposits account for 77.3 per cent of the 8.37×10^9 tons of coal accredited to the Lake Baikal region. However, over seventy deposits are known to exist in this area. Most are small and dispersed over a wide range of territory. Shabarova and Tyshnova identify eighteen of these on their map, and Troyanski depicts thirty-eight. Five $\times 10^9$ tons of these reserves supposedly are hard coals, but since they are of dubious quality they cannot be placed within any of the established hard coal categories. Thiel relegates all to a brown coal classification. [39] In the past two decades, estimates of geological reserves increased by 290 per cent. They now comprise 0.1 per cent of the Soviet Union's total reserves.

Quality and amounts of the seven major deposits are presented in the following listing:

Name of Deposit	Grade of Coal	Reserves[a] in Tons 10^6	Moisture in[b] Air Dried Coal %	Ash in[b] Dry Coal %	Sulfur in[b] Dry Coal %
Lake Gusin	B	3,456	5.0	15 - 20	--
Tarbagatay	B	170	8.0	12 - 25	2.5 - 4.0
Chernov	B	66	11.0	10 - 19	--- - 1.0
Kharanor	B	1,769	21.0	8 - 14	--- - 1.0
Arbagaro-Kholbon	B	121	11.0	-- - 15	--- - 2.0
Tugnuy[c]	D-G	612	---	-- - 20	--- - 1.0
Bukachacha	D-G	47	11.0	-- - 12	--- - 1.0

[a]Shabarova and Tyshnova, *op. cit.*, p. 139.

[b]Troyanski, *op. cit.*, pp. 392, 393, 394.

[c]Anon., "New Hard Coal Basin," *Geografiya V Shkole*, No. 5 (1957), p. 66. This article implies that the quality is higher than stated here but does not present qualitative data.

These brown coal deposits are superior to any of the European or Ural deposits. Except for a slightly higher moisture content the hard coals are equal to the poorer Pechora coals of a similar grade.

The Soviet Far East.—Far Eastern deposits, including the Island of Sakhalin have 0.6 per cent or 55.7×10^9 tons of the geological reserves of Soviet coal. Hard coal constitutes 43.8×10^9 tons and brown coal 11.9×10^9 tons. Seven of the thirty-eight basins and deposits contain 33×10^9 tons of the reserves.

Bureya is the largest hard coal basin in the Far East. Geological reserves total 25×10^9 tons, and of this 24×10^9 tons are mineable.

This coal has the same volatile and carbon content as comparable Donets coals; its sulfur content is low, but it contains a great deal of ash, the maximum reaching 48 per cent. These coals are not concentrated before shipment, and as a result have the highest ash content in the U.S.S.R. (see Table 2).

Other exclusively hard coal basins are:

Name of Deposit	Grade of Coal	Reserves in Tons 10^9	Moisture in Air Dried Coal %	Ash in Dry Coal %	Sulfur in Dry Coal %
Suifun	D	1.66	0.6 - 2.0	12 - 27	-- - 0.4
Suchan	A, T, K Pzh	1.43	-- - 5.0	21 - 27	-- - 1.0

Low in sulfur content, these coals rank higher than similar grades found in the Donets Basin, and their heating capacity is greater. Their importance is increased by their location, near the city of Vladivostok.

Significant Far Eastern brown coal deposits are located at:

Name of Deposit	Reserves in Tons 10^6	Moisture in Air Dried Coal %	Ash in Dry Coal %	Sulfur in Dry Coal %
Bikinsk[a]	2,900	19	22 - 35	-- - 0.6
Uglov (Artem)	1,020	16	25 - 35	-- - 0.6
Mikhinsk	60	-	4 - 40	0.1 - 0.6
Suputinsk	45	16	22 - 40	-- - 1.0

[a] V. G. Udovenko, *The Far East, Economic-Geographic Characteristics* (Moskva: Geografgiz, 1957), p. 81. This author gives the ash content of Artem coal as 19 per cent.

Brown coal deposits of the Soviet Far East are the least contaminated with sulfur of any of the Soviet brown coals. Ash and moisture contents are less than Dneper and Moscow coals, and calorie contents are greater. After concentration, the ash content does not exceed 19 per cent. [40]

Sakhalin Island has 20×10^9 tons of coal distributed throughout 8 deposits. Fifteen billion tons are hard coal and 5×10^9 tons brown coal. Nineteen billion tons are mineable. Both brown and hard coal are low in sulfur and high in ash. Brown coals are high in volatiles and carbon, and have a greater heating capacity than their European counterparts. Hard coals are superior in carbon and heating capacity to similar Donets coal.

The Northeastern Territory.—The 239.9×10^9 tons of coal located in the Northeastern Territory constitute the remaining 2.8 per cent of Soviet geological reserves. These reserves distributed throughout 28 basins and deposits represent an increase of 135 per cent over the 1937 estimates. Fifty-two per cent of the geological reserves are brown coal. Mineable reserves of all types of coal total 184×10^9 tons.

Zyryansk, located in the Indigrka-Kolyma interfluve, has 43 per cent of the total geological reserves and 73 per cent of the hard coal reserves. Zyryansk coal compares favorably in quality with Donets and Kuznets coals. Its moisture, ash, and sulfur contents are lower; however, its volatile content is higher, and carbon content lower.

Data on the quality of other hard coals in the North East are unavailable except at the Bukhta-Ugolnaya Basin. Bukhta's 6.18×10^9 tons of coal are similar to the hard coals of Transbaikalia. They are high in moisture, ash, and volatile.

Anadyr possesses 78 per cent of the Northeast's brown coal. Qualitative data on the 97.91×10^9 tons of reserves in this deposit and other brown coal deposits of the Northeast are unavailable.

Kamchatka has 830×10^6 tons of hard coal in ten known deposits on the Peninsula. Data concerning its quality are also unavailable.

Correlative Summary on Reserves and Quality.—The enormity of the Soviet claim to coal reserves demands evaluation rather than "prima facie" acceptance. Such an evaluation does not imply that Soviet statistics are incorrect, nor that the technical personnel originally making the surveys lacked competence. A true understanding of this claim does require that it be interpreted in terms of our own standards for measuring coal reserves and their quality. Several differences exist; namely depth to which reserves are computed, thickness of the seams included in the estimate, and quality of the coal.

Coal reserves are traditionally estimated to a depth of 1,800 meters in the Soviet Union. Reserves of coal in other countries are seldom estimated below 1,000 meters, and brown coal or lignite is not estimated below 500 meters in the United States. Comparative data for exactly the same depths are not available, but if Soviet hard coal is considered to a depth of only 1,200 meters, a depth which is similar to that for other countries, reserves total but 4.4×10^{12} tons. Brown coal measured to a depth of 500 meters in the Soviet Union totals just 1.0×10^{12} tons. Reserves then total 5.4×10^{12} tons rather than the 8.6×10^{12} tons claimed. Computations of Soviet reserves on depth standards comparable to those used in Western countries would still permit the Soviets to claim first place in reserves throughout the world. It must be understood that depth limits of 1,000 or 1,800 meters are geographically and economically determined limits, not geologically determined limits, and in the case of the United States and other Western nations, economic and geographic conditions are greatly different than they are in the Soviet Union. Certain geographic exigencies may make it imperative for the Soviets to mine coal at depths exceeding 1,000 meters. This is true in European Russia today; Donbas mines are being planned below the 1,000 meter level because coking coals in the upper levels are becoming exhausted. [41] The alternative to this deep mining would mean bringing coking coals into European Russia in vast quantities from Arctic or Siberian Basins. The distances and shipping problems involved would be tremendous.

Actual differences in the thickness of coal seams included in reserve estimates in the United States and the Soviet Union are slight. With the exception of the Donets Basin, coal seams in the Soviet Union must attain a thickness of approximately 15.6 inches before they are included in estimates of reserves. Current United States criteria consider seams of anthracite, semianthracite, and bituminous coals suitable if they have a thickness of 14 to 28 inches. [42] Subbituminous and lignite seams are included in estimates if they are $2\frac{1}{2}$ to 5 feet thick. Brown coal in the U.S.S.R. is included in the estimates if the seams are 16.5 inches in thickness. By incorporating thinner seams of brown coal, the Soviet naturally will be able to increase the total volume of their reserves. Standards of thickness of seams are different for the Donets Basin. In this basin, seams with a thickness of 11.7 to 15.6 inches are included in the reserve estimate. By including this thin seam coal, the Soviets materially add to their total reserve fund. However, the dictates of geography may make the mining of this coal feasible in the future.

The quality of coal included in estimates of reserves differs substantially between the United States and the Soviet Union. In the latter, coal with an ash content of 50 per cent is included, and in the Moscow Basin, the permissive ash limit is 60 per cent. According to Averitt and Berryhill, coal with an ash content exceeding 33 per cent is not included in reserve estimates in the United States. Once again the geography of the Soviet Union indicates that the material at hand must be utilized because the distances involved in shipping coal in such a vast country are great. Chukhanov, in his analysis of Moscow and Donets coals in power plants in the Moscow region, has proven that electric power generated with Moscow coal costs 20 to 60 per cent more than power generated from Donets coal. [43] Obviously this is part of the justification for mining thinner and deeper seams in the Donets Basin.

Although there is justification for critically evaluating the methodology which the Soviets have applied in computing their reserves, there is no justification for rejecting these claims within the frame of reference in which they were made. If, in time, these estimates prove inaccurate, it will be to the detriment of Soviet regional planning. Because the functioning Soviet economy is a product of planning by geographic region, a summary of the spatial aspects of these reserves is dictated.

In terms of quality and reserve tonnage, basins and deposits located east of the Urals have over 90 per cent of the Soviet Union's coal. Among the industrially developed basins, the Kuznetsk ranks first in quality. European Russia, excluding the Urals, has but 7.5 per cent of the reserves. They are, in general, excellent to very poor coals. The Pechora Basin, in terms of geological and mineable reserves, and quality, is the outstanding basin in European Russia. The Urals and the Caucasus are not only low in reserves, but possess low quality coals. Middle Asia contains coals of all quality but is low in reserves when compared to regions to the north and east. Kazakhstan, with its famed Karaganda Basin, possesses high quality and very poor quality coals. Karaganda itself is one of the high quality regions of the U.S.S.R. Transbaikalia, one of the older mining regions of the country, has extremely low quality coal and few reserves. Recent surveys make it possible to credit the Arctic and Subarctic Region and the Northeast with not only the greatest reserves of the country, but with some of the highest quality ones.

Untouched and undeveloped in Soviet Russia are vast reserves of high quality coals, located at the present stage of technological development in isolated and remote regions. More intensive survey work may result in decreasing the tonnage accredited to the probable and possible categories, but it will undoubtedly add to the proven category. Soviet Russia does not lack coal reserves of high quality; she lacks the facilities for using these reserves, and in that respect, her geography is hostile. Distance and the transportation facilities needed to overcome it are her main enemies.

THE ENERGY POTENTIAL OF SOVIET RUSSIA'S COAL RESERVES

The Total Energy of Soviet Coal Reserves.—Current geological reserves of coal, when measured in terms of their energy content, contain a potential 54.7×10^{15} kilowatt hours, 274×10^{6} kilowatt hours per capita (see Table 3). This represents an increase of 41.4×10^{15} kilowatt hours in energy potential since 1937. Mineable reserves, which are a more accurate measure of potential, contain 48.1×10^{15} kilowatt hours, with a per capita potential of 241×10^{6} kilowatt hours. Hard coal possesses 83.0 per cent of this potential and brown coal 17.0 per cent.

In 1955, the Soviet Union generated 170×10^{9} kilowatt hours of electricity; 64.0 per cent [44] of this energy was produced by only 25 per cent of the coal mined in that year. [45] Per capita production at the time was 861 kilowatt hours; [46] coal produced 545 of these kilowatt hours. [47] In the same year, France had a per capita production of 1,132 kilowatt hours; England, 1,707; and the United States, 3,782. [48] Soviet production of electroenergy is low when compared

TABLE 2

THE ASH CONTENT OF COAL SHIPPED TO CONSUMERS IN PER CENT, AND COAL DRESSED IN THOUSANDS OF TONS (10^3), BY THE MINISTRY OF THE COAL INDUSTRY, SELECTED YEARS 1940-1955[a]

Basin or Combinat	1940	1945	1950	1955
Donets				
Ash content	13.1	17.4	14.1	14.5[b]
Coal dressed	9,175.0	538.1	10,511.4	36,405.1
Pechora				
Ash content	----	----	----	19.2
Coal dressed	----	----	----	3,567.6
Kizel				
Ash content	25.2	25.4	25.1	24.7
Coal dressed	----	----	----	456.4
Karaganda				
Ash content	18.5	18.7	20.2	22.1
Coal dressed	662.5	1,721.6	2,006.7	6,712.4
Kuznetsk				
Ash content	10.5	9.2	9.7	10.5
Coal dressed	----	4,725.0	11,342.8	20,522.6
Combinat Vostsibugol				
Ash content	15.3	13.9	14.3	17.2
Coal dressed	----	----	----	2,150.0
Far East Deposits				
Ash content	29.6	27.5	21.1	22.5
Coal dressed	63.3	70.8	176.4	1,324.0
Combinat Primorskugol				
Ash content	29.6	27.5	27.2	27.5
Coal dressed	63.3	70.8	90.0	124.4
Combinat Far Eastugol				
Ash content	----	----	33.0	32.3
Coal dressed	----	----	----	----
Combinat Sakhalinugol				
Ash content	----	----	12.8	14.2
Coal dressed	----	----	886.4	1,199.6
Georgian Deposits				
Ash content	30.4	29.9	28.2	29.5
Coal dressed	----	----	770.2	2,200.5
Middle Asia				
Ash content	18.0	19.8	16.3	14.9
Coal dressed	----	----	----	----

[a] Compiled from A. G. Perdukhin, ed., *Ugolnaya Promyshlennost SSSR Statisticheskiy Sbornik* (Moskva: Ugletekhizdat, 1957), pp. 78-79 and 83-84.

[b] D. T. Onika, *Ugolnaya Promyshlennost SSSR V Shestoi Pyatileke* (Moskva: Ugletekhizdat, 1956), p. 39. This author lists the 1955 ash contents given here as averages for the basins.

with other countries, but her ultimate potential is great. Were the Soviets entirely dependent on coal as a source of energy, production could still be expanded because of the tremendous reserves. If it were possible to use all of the energy contained in the mineable reserves of coal, the Soviets would have enough electricity to refine 2.9×10^{15} tons of oil, or produce 430×10^{12} tons of rolled steel, or produce 253×10^9 tons of aluminum. [49] Large amounts of energy,

TABLE 3

THE POTENTIAL KILOWATT HOUR CONTENT OF GEOLOGICAL AND MINEABLE RESERVES OF SOVIET COAL, IN MEGA KILOWATT HOURS, 1937-1956

Years and Category	Energy Potential Total Coal	Energy Potential Hard Coal	Energy Potential Brown Coal
	Mega Kilowatt Hours		
1937 geol. reserves[a]	12,328,799,800	11,745,442,800	583,337,000
1956 geol. reserves[b]	54,794,448,200	46,035,717,200	8,758,771,000
1937-1956 increase[c]	41,465,688,400	33,290,274,400	8,175,394,000
1956 mineable reserves[d]	48,233,118,420	39,912,772,778	8,320,345,632
	Mega Kilowatt Hours per Capita		
1936 geol. reserves[e]	72	69	3
1956 geol. reserves[f]	274	230	44
1937-1956 increase[g]	202	162	40
1956 mineable reserves[h]	241	199	42

[a]Computed from M. M. Prigorovsky, "Coal Bearing Provinces and Basins of the U.S.S.R.," *Report of the XVII Session, International Geological Congress*, Vol. I. (Moscow: 1939), p. 194.

[b]Computed from data in Coal Appendix – Table I.

[c]Computed from (a) and (b) above.

[d]Adapted from Coal Appendix – Table II.

[e]Computed from data in (a) above and population data presented in TS. S.U. S.S.S.R. (S. Ya. Genin, ed.), *National Economy of the U.S.S.R., Statistical Manual* (Moskva: Gosstatizdat, 1956) p. 17, population 1939.

[f]Computed from (b) above and population data in S. Ya. Genin above.

[g]Computed from (c) above and population data in S. Ya. Genin above, p. 17, estimated population U.S.S.R., April 1956.

[h]Computed from Coal Appendix – Table II and S. Ya. Genin, p. 17, estimated population U.S.S.R., April 1956.

such as that inherent in Soviet coal resources, are impressive, but the significance of this enormous energy potential is dependent entirely upon its geographic distribution.

GEOGRAPHICAL DISTRIBUTION OF THE ENERGY POTENTIAL OF SOVIET COAL RESERVES

Energy potentials of Soviet coal reserves basically reflect the same distributional pattern as that of the geological and mineable reserves (see Figure 3). The following listing of energy potentials by traditionally established regions reveals that the area east of the Urals contains 91.6 per cent of the total coal energy potential of the Soviet Union and that the Urals, Caucasus, and European Russia contain but 8.4 per cent.

Traditionally Established Regions	Per Cent of Mineable Reserves	Energy Content of Mineable Reserves Kwh. 10^{12}	Per Cent of Energy Content
European Russia	6.52	4,004.9	8.30
Caucasia	0.01	7.5	0.02
Urals	0.09	28.4	0.06
Kazakhstan	1.58	716.8	1.49
Middle Asia	0.49	239.9	0.50
Arctic and Subarctic Siberia	60.52	30,674.5	63.59
South Siberian Belt	27.61	11,109.9	23.03
Transbiakal	0.09	42.3	0.09
Sakhalin	0.21	130.1	0.27
Far East	0.50	254.2	0.53
North East	2.38	1,023.5	2.12
Totals	100.00	48,233.0	100.00

Isolated areas, such as Arctic and Subarctic Siberia and the North East, possess approximately two thirds of the total energy potential.

Where variations exist between the per cent of the mineable reserves and the per cent of the energy content of the mineable reserves, they are the result of differing proportions of brown coal and hard coal in a region. A comparison of the seven basins which contain 90.4 per cent of the mineable reserves illustrates the significance of this variation. The Kansk-Achinsk brown coal basin, with approximately 16 per cent of the reserves, has but 7 per cent of the energy potential, whereas the Taimyr hard coal basin, with only 7 per cent of the mineable reserves, has 8 per cent of the energy potential. The Lena Basin, which is predominantly brown coal, has nearly a third of the mineable reserves but only a fourth of the energy potential of the country.

Basin	Per Cent of Mineable Reserves	Per Cent of Energy Content
Lena	32.4	25.1
Tunguska	19.5	25.6
Kansk-Achinsk	15.6	7.3
Kuznetsk	10.4	13.6
Taimyr	6.6	8.3
Pechora	3.5	4.4
Donets	2.4	3.2
Totals	90.4	87.5

While these figures broadly outline the framework of the distribution of energy potentials to be derived from coal, it is only by a consideration of the regions themselves that the true geographic importance is realized.

European Russia.—Two hard coal basins, the Donets in the extreme

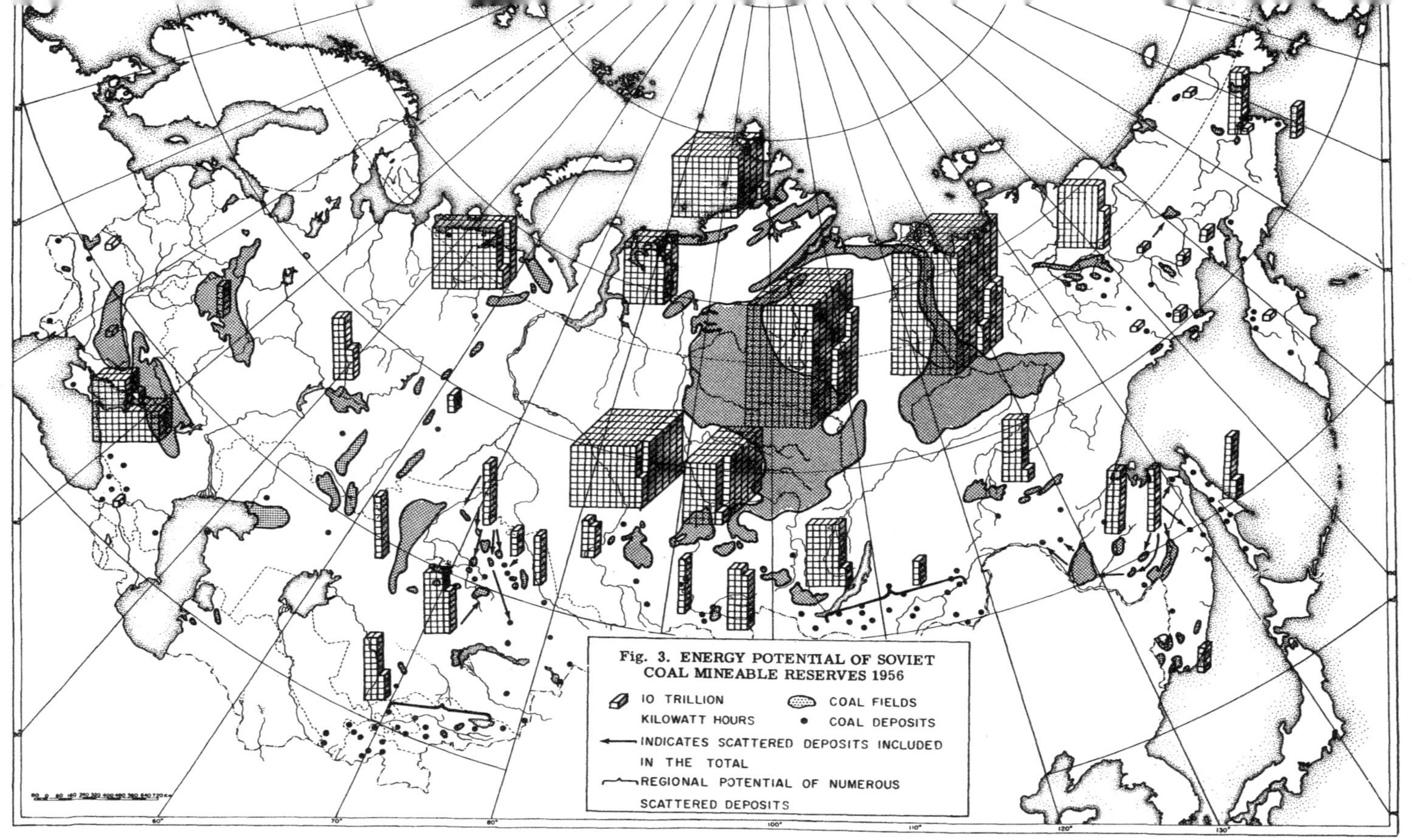

Fig. 3. ENERGY POTENTIAL OF SOVIET COAL MINEABLE RESERVES 1956

South and the Pechora in the extreme North account for 91.9 per cent of the 4×10^{15} kilowatt hours of European energy potential. The industrially developed Dontes has 38.6 per cent of the total European potential and the relatively unexploited Pechora has 63.3 per cent. The Moscow brown coal basin, which is intensively utilized industrially because it represents the only energy resource in the Central Region, possesses but 1.26 per cent of the energy potential of European Russia.

The Dneper brown coal basin in the Ukraine [50] is becoming more significant locally. Its 3.6×10^9 tons of mineable reserves contain 0.27 per cent of the energy potential. Equal in energy potential to the Dneper Basin, is the newly discovered Lvov-Volyn hard coal basin; its 1.4×10^9 tons of mineable reserves contain 0.29 per cent of the European energy potential. Because of the higher energy values, coal from this basin is destined for wider use in the West European Part of the U.S.S.R. [51] Strategically located in relation to the Volga Region, is the Kama hard coal basin with 5.89 per cent of the European energy potential. While favored by location, it is somewhat condemned by its high sulfur content, and the depth of the coal bearing strata (see Coal Appendix Tables - 1 and 3).

Caucasia.—This region with but 7.5×10^9 kilowatt hours has the lowest coal energy potential of any coal producing region in the U.S.S.R. Mineable reserves and the energy potential to be derived from them are only 0.02 per cent of the country's total. In this region, 95.0 per cent of the energy potential emanated from hard coal. North slope hard coal deposits possess 36.7 per cent of the potential and the balance is found in the Georgian S.S.R.

The Urals.—The industrially important Urals possesses only 0.06 per cent, or 28.3×10^6 kilowatt hours, of the total energy potential of Soviet coal reserves. Brown coal predominates, constituting 77.5 per cent of the mineable reserves and 55.0 per cent of the energy potential.

The Kizel hard coal basin, with its highly sulfurous and deeply bedded strata, has 30.5 per cent of the energy potential of the Ural Region. Another 14.3 per cent of the energy potential is located in the hard coal deposits in the vicinity of Dombarovski.

Four widely dispersed brown coal basins account for 14.5×10^{12} kilowatt hours of the total 15.6×10^{12} kilowatt hour energy potential possible from brown coal. They are the South Ural Basin, with 15.0 per cent of the potential; the Chelyabinsk Basin, with 15.5 per cent; the Orsk East Ural Basin with 10.2 per cent; and the North Sosvinsk Coal Region, with 10.8 per cent of the potential. Each of these regions is the site of at least one major "Thermoelectric Station" in the Ural grid. [52] The remaining 3.9 per cent of the potential is in small scattered brown coal deposits.

Kazakhstan.—The 1.58 per cent of the Soviet mineable reserves located in Kazakhstan have an energy potential of 716 × 10^{12} kilowatt hours or 1.48 per cent of the total energy potential of the country. Five of the thirty-one basins and deposits in this region contain 85.0 per cent of the energy potential.

Basin	Per Cent of the Region's Energy Potential
Karaganda	52.16
Ubagan	14.36
Ekibastuz	12.23
Maikyubensk	5.44
Lenger	0.81
	85.00

Karaganda and its satellite, Ekibastuz, have approximately 65.0 per cent of the total energy potential of Kazakhstan, but only 46.7 per cent of the mineable coal in the Republic. The advantage which accrues to these areas in energy potential is derived from hard coal.

Three brown coal deposits, the Lenger, the Maikyubensk, and the recently discovered Ubagansk basin, have approximately 20.0 per cent of the energy potential. Although they contain 41.5 per cent of the mineable reserves, their energy potential is low because of the lower thermal values of brown coal. Located in areas rich in nonferrous minerals, which have high electricity requirements for reduction, these brown coals may be used in thermoelectric plants as a source of power in the metals and chemical industry.

The remaining 15.0 per cent of the energy potential is derived from 11.8 per cent of the mineable reserves, scattered throughout twenty-six hard and brown coal deposits.

Middle Asia.—Middle Asia contains less coal than the newly discovered South Yakutsk Basin. Mountainous terrain and vast stretches of desert separate most of the twenty-two deposits which together contain but 0.5 per cent of the energy potential of the Soviet Union. It is only in the Fergana Valley that any significant concentration of the energy potential is located. Of the 239 × 10^{12} kilowatt hour potential credited to Middle Asia, approximately 22.5 per cent is located in the following four deposits in the Fergana Valley; East Fergana (Uzgen), 10.3 per cent; North Fergana, 8.6 per cent; Kyzyl-Kiya, 2.8 per cent; and the Sulyukta deposit, 0.8 per cent. [53]

Even though the energy potentials of the individual deposits in this region are infinitesimal, they do have importance as local fuels. One hundred kilometers from Tashkent is the Angren brown coal deposit, which constitutes but 3.4 per cent of the energy potential. It is the site of the largest underground coal gasification plant in the

U.S.S.R. [54] This deposit and the type of development being carried on in it may be indicative of the future method of working all small Middle Asian coal deposits.

Hard coal deposits in Middle Asia contain 83.4 per cent of the energy potential and brown coal 16.6 per cent.

Arctic and Subarctic Siberia.—Isolated and undeveloped Arctic and Subarctic Siberia has approximately 64.0 per cent of the coal energy potential of the Soviet Union. Mineable reserves are made up of 70.0 per cent hard coal and 30.0 per cent brown coal. However, 86.0 per cent of the region's energy potential is contained in the hard coal and 14.0 per cent in the brown coal.

Five basins account for practically all of the 30.6×10^{15} kilowatt hour energy potential. They are, in the order of their magnitude:

Basin	Per Cent of the Region's Energy Potential
Tunguska	40.23
Lena	39.94
Taimyr	13.09
Ust-Yenisey	5.68
South Yakutsk	1.05
Other	0.01
	100.00

The smallest of these five major basins, the South Yakutsk, has an energy potential several times greater than the Urals and Caucasia combined. The Lena and the Tunguska Basins have over 80 per cent of the energy potential of this region and over 50 per cent of the total energy potential of the Soviet Union. The Tunguska Basin, alone, has three times the combined energy potential of the Pechora and Donets Basins, and more than twice the energy potential of the Kuznetsk Basin. Taimyr, the third largest basin in the region, has an energy potential equal to all of European Russia.

The South Siberian Belt.—Mineable reserves of this region are nearly equal to the total geological reserves of the United States. [55] The 11.1×10^{15} kilowatt hour energy potential of these mineable reserves constitutes 23.0 per cent of the total energy potential of Soviet Russia's coal. Hard coal is responsible for 68.3 per cent of this potential and brown coal 31.7 per cent.

Seven basins, as listed at the top of the facing page, with the famous Kuznetsk ranking first among them, have the vast preponderance of the coal energy potential.

Within the Kuznetsk Basin, there is a greater energy potential than all of the following regions: European Russia, Caucasia, Urals, Kazakhstan, Middle Asia, Transbaikal, the Far East, Sakhalin, and the North East. Two of the largest thermoelectric stations in the

Basin	Per Cent of the Region's Energy Potential
Kuznetsk	58.92
Kansk-Achinsk	31.69
Irkutsk	4.77
Minusinsk	2.66
Gorlov	1.10
Tuvinian	0.79
Tomsk	0.06
Other	0.01
	100.00

U.S.S.R. are being constructed in the Seventh Five Year Plan to utilize this fuel. [56]

Kansk-Achinsk, the largest basin in the region in terms of tonnage reserves, has an energy potential of 3.5×10^{15} kilowatt hours. Its 31.7 per cent of the region's energy potential is derived from brown coal. Forty per cent of the Soviet Union's brown coal is located in this basin; this is equivalent to 15.6 per cent of the total mineable reserves, and 7.3 per cent of the total energy potential.

Minusinsk, a hard coal basin in the Khakass Autonomous Oblast, has 2.66 per cent of the energy potential, an amount which is greater than all of Middle Asia or the Far East.

Adjacent and west of the Kuznetsk is the Gorlov Basin with 1.10 per cent of the energy potential. Entirely composed of anthracite coal, this small basin has an energy potential three times larger than both the Caucasian and Urals regions.

Hard coal deposits of the Tuvinian Autonomous Oblast have only 0.79 per cent of the energy potential, but this is more than twice the potential of the entire Transbaikal region.

The balance of the energy potential is based upon brown coal, primarily deposits in the Tomsk Oblast.

Transbaikal Region.—Transbaikalia, referred to by the Soviets as one of the "old mining industrial centers," possesses 0.09 per cent of the energy potential of the Soviet Union's coal reserves. [57] Its 42.3×10^{12} kilowatt hours of energy potential is dispersed in seven small basins and scores of smaller deposits. Two brown coal basins, the Lake Gusin and the Kharanorsk, have nearly fifty per cent of the energy potential of Transbaikal. Preference in development is to be given to the smaller of these two basins, the Kharanorsk, probably because its coal has a lower ash content and higher thermal value. [58]

The Soviet Far East.—Far Eastern deposits comprise but 0.53 per cent of the Soviet Union's total coal energy potential. The Bureya Basin dominates with 79.87 per cent of the 254.2×10^{12} kilowatt hour energy potential.

A secondary concentration occurs in the Primorski Krae north of Vladivostok. Four basins with established mining areas have 14.18 per cent of the potential.

Basin	Per Cent of the Region's Energy Potential
Bikinsk	3.32
Suifun	5.19
Suchan	4.49
Uglov (Artem)	1.18
	14.18

An additional six per cent of the potential is located in numerous small deposits. Most of these deposits are unexploited to date.

Coal in this region has a greater energy potential than Middle Asia and Caucasia. Much of the coal mined is used for power on the Transiberian Railroad, and in thermoelectric stations at Khabarovsk, Komsomolsk, Suchan, and the Atrem-Vladivostok station. [59]

Sakhalin Island.—Remote Sakhalin Island possesses 0.27 per cent of the total coal energy potential of the Soviet Union. The 130.1×10^{12} kilowatt hours located in the coal of this island exceed by three and one-half times the coal energy potential of the Urals and Caucasian regions. Eighty-eight per cent of the potential is based on hard coal and 12 per cent on brown coal.

The Northeastern Territory.—Energy potentials of the Soviet North East are superior to the total potentials of the Caucasus, Urals, Kazakhstan, and Middle Asia. Located in this region, which is even more isolated and remote from established industrial centers than Arctic Siberia, is a potential $1{,}023.3 \times 10^{12}$ kilowatt hours of coal energy, or 2.2 per cent of the Soviet's total. Fifty-one per cent of the mineable reserves are hard coal and 49 per cent brown coal; however, 75 per cent of the energy potential is derived from hard coal and 25 per cent from brown coal.

Bisected by the Arctic Circle, the Zyryansk hard coal basin contains 65 per cent of the region's energy potential. Another five per cent of the potential is located in the Bukhta hard coal basin south of the Pacific port city of Anadyr. North and west of this city is the Anadyr brown coal basin with 20.2 per cent of the potential. West and east of Magadan is the Okhotsk brown coal area, with an additional 2.4 per cent of the potential. Scattered deposits have the remaining seven per cent of the region's potential. Kamchatka has about one per cent.

CONCLUSIONS ON THE DISTRIBUTION OF ENERGY POTENTIAL

Soviet Russia has the greatest coal energy potential of any nation in the world provided that the estimates of reserves are correct. Over 90 per cent of this potential is located east of the Urals and north of Middle Asia, with the vast preponderance in presently inaccessible Arctic and Subarctic regions of Siberia or the North East. The small percentage of the energy potential allocated to European Russia, including the Urals and Caucasia, would be even less were it not for the fact that mineable reserves have been computed for greater depths in this region (Donets and Kizel Basins), and the Moscow Basin includes material classified as coal which is inadmissible by Soviet standards in other regions. Regions with a high energy potential correspond to the regions of high quality coal, and with the distribution of geological and mineable reserves.

A complete realization of the significance of Soviet Russia's energy potential is possible only when viewed within the framework of production and consumption.

NOTES – CHAPTER I

1. Frederick Brown, *Statistical Yearbook of the World Power Conference No. 4* (London: Central Office, World Power Conference 1948).

2. Paul Averitt and Louise R. Berryhill, "Coal Resources of the United States," *Geological Survey Circular 94*, (1950), 26.

3. Akademii Nauk S.S.S.R., "Materials on the Coal Geology Conference," *Trudy Institute Geologicheskikh Nauk, No. 90. Coal Series* No. 12 (1947), 227.

4. P. Antropov, Minister of Geology and Conservation of Minerals, "Prirodnye Bogatstva Strany Na Sluzhby Narodnomy Khozyaistvu," *Pravda*, No. 338 (14387), 4 Dec. 1957, p. 2.

5. N. V. Shabarova and A. V. Tyzhnova, ed., *Zapasy Ugley i Goryuchikh Slantsy S.S.S.R.* (Moskva: Gosgeoltekhizdat, 1958), p. 30.

6. Elwood S. Moore, *Coal: its Properties, Analysis, Classification, Geology, Extraction, Uses and Distribution* (New York: John Wiley and Sons, Inc., 1940), pp. 113-135. Moore lists twelve classifications in use in the United States, Canada, and England.

7. A. G. Betekhtin, *Kurs Mestorozhdenii Poleznykh Iskopayemky* (Moskva: Gostopekhizdat, 1946), p. 424.

8. E. S. Moore, *op. cit.*, p. 125.

9. A. A. Gapeev, *Tverdyye Goryuchiye Iskopayemyye* (Moskva: Goa. Izd-Vo. Geologicheskoi Lit-ry, 1949), p. 21. This author lists the ash limit as 40 per cent. N. V. Shabarova and A. V. Tyshnova, *op. cit.*, p. 30. These authors list the ash limit as 50 per cent.

10. Coal Appendix – Table I and II lists geological and mineable reserves.

11. A geological map of recent discoveries is found in S. V. Troyanski, ed., *Gornoe Delo Entskilopedicheskiy Spravochnik, Vol. 2* (Moskva: Ugletekhizdat, 1957), p. 196-197 insert.

12. E. P. Shevchenko, *Ugli Dlya Koksovaniya*, (Moskva: Ugletekhizdat, 1944), p. 24.

13. V. P. Aksenov, N. P. Zamorenov, and A. D. Rybin, *Razrabotka Burykh Uglei Ukrainy* (Kiev: Gosudaretvennoe Izdatelstvo Tekhnicheskoi Literatury, 1944), p. 13.

14. M. B. Ravich, *Toplivo V Shestoy Pyatileka* (Moskva: Gospolitizdat, 1956), p. 19.

15. *Ibid.*, p. 18.

16. N. V. Shabarova and A. V. Tyshova, *op. cit.*, p. 30.

17. *Pravda*, No. 234 (13897), 21 August 1956, Dateline Novovolynsk, p. 1.

18. A. T. Vasrchenko, "Lvovsko-Volynskoly Ugol'nyy Bassein," *Geografiya V Shkole*, No. 1 (1956), p. 14.

19. N. V. Shabarova and A. V. Tyshnova, *op. cit.*, p. 69.

20. A. L. Odud, *Moldavskaya S.S.R.* (Moskva: Geografgiz, 1955), p. 15.

21. E. P. Shevchenko, *op. cit.*, p. 29.

22. E. P. Maslov, A. I. Gozulov, and S. N. Ryazantsev, eds., *Severyy Kavkaz*, (Moskva: Geografgiz, 1957), p. 94. These writers claim that 81 per cent of the hard coal in North Slope deposits is anthracite; analysis indicates otherwise (see Coal Appendix – Table III).

23. B. I. Kustov and G. L. Kushev, "Ugli Kazakhstana Ikh Kachested i Klassifikatsiya," *Trudy Instituta Geologicheskikh Nauk*, No. 90 (1947), 149. These authors list 360 manifestations of coal in Kazakhstan but dismiss most as insignificant.

24. A history of this survey appears in I. I. Satpasv, *Osnovnyitogi Geologicheskogo Izhucheniya Karagandinskogo Basseina Za 25 Let* (Alma-Ata: Akademii Nauk Kazakhskoy S.S.R., 1956), pp. 10-14.

25. Kustov and Kushev, *op. cit.*, p. 152.

26. S. Batishchev-Tarosov, "Bogatstva Turgaiskoy Stepi-Na Sluzhby Rodine," *Pravda*, No. 148 (13811), 27 Mar. 1956, p. 2.

27. Discovery of an additional 100 million tons of hard coal at Tyulek-Markaiski was reported in *Pravda*, No. 17 (14411), 17 Jan. 1958, p. 1.

28. Shabarova and Tyshnova, *op. cit.*, p. 94.

29. Gapeev, *op. cit.*, p. 309.

30. Troyanski, *op. cit.*, p. 398.

31. Only 1.1 x 10^{12} tons of coal are credited to this basin by the following: B. V. Tkachenko, M. I. Rabkin, K. K. Demokidov, V. A. Vakar, A. L. Grozdilov, E. L. Butakov and S. A. Strelkov, "Geologicheskoe Stroenie Severnoy Chasti Sredne-Sibirskogo Ploskogorya," *Geologiya Sovetskoy Arktiki* (Moskva: Gosgeoltekhizdat, 1957), p. 240.

32. Anon. *Geografiya V Shkole*, No. 3, 1958, p. 62.

33. *Voprosy Geologii Kuzbass*, No. 1 (Moskva: Ugletekhizdat, 1956), 248 pp. This issue outlines the purpose of the series and deals with stratigraphic problems.

34. A. Zademidko, "Rezervy Ekonomicheskikh Rainov-Na Sluzhby, *Pravda*, No. 203 (14232), 22 July 1957, p. 2.

35. Shevchenko, *op. cit.*, p. 28.

36. *Ibid.*

37. V. A. Annenkov, G. N. Dmitriev, K. I. Syskov, and A. N. Strukov, "Metallurgicheskiy Koks Iz Ugley Irkutskogo Basseina," *Vostochnykh Filialov Akademii Nauk*, Novosibirsk, No. 6 (1957), 77. Same authors, "Metallurgical Coke from Coals of the Irkutsk-Cheremkovo Basin," *Izvestia Akademii Nauk, S.S.S.R. Otdelenie Tekhnicheskikh Nauk*, No. 7 (1958), 113.

38. L. Semenov, "Ugolnye Resursy Krasnoyarskogo Kraya i Perspektivy Ikh Ispolzovaniya," *Planovoe Khozyaistvo*, No. 12 (Dec., 1957), p. 77.

39. Erich Thiel, *The Soviet Far East, A Survey of Its Physical and Economic Geography* (London: Methuen and Co., Ltd., 1957), p. 163.

40. A. G. Perdukhin, ed., *Ugolnaya Promyshelnnost S.S.S.R., Statisticheskiy Spravochnik* (Moskva: Ugletekhizdat, 1956), p. 85.

41. B. I. Andreev, D. V. Kravchenko, *Kamennougolnye Basseiny S.S.S.R.* (Moskva: Gosudarstvennoe Uchebno - Pedagogicheskoe Izdatelstvo Ministerstva Prosveshcheniya R.S.F.S.R., 1958), p. 57.

42. Averitt and Berryhill, *op. cit.*, p. 9.

43. Z. Chukhanov, "The Moscow Coal Basin: Concerning the Effectiveness of Its Development," *Problems of Economics*, Vol. 1, No. 9 (January, 1959), p. 23. Translated contents of January 1959 Voprosy Ekonomiki, International Arts and Sciences Press.

44. Ts. S.U. S.S.S.R., (K. G. Ivanov, ed.), *Promyshlennost S.S.S.R. Statisticheskiy Sbornik* (Moskva: Gosstatizdat, 1957) p. 178.

45. L. Kudryashov, "Rezervy Ekonomii Topliva Na Elektrostantsiykh," *Planovoe Khozyaistvo*, No. 7 (1957), p. 74.

46. I. A. Kulev, *Elektrifikatsiia S.S.S.R. V Shestoi Piatiletke* (Moskva: Gospolitizdat, 1957), p. 8.

47. Ts. S.U. S.S.S.R., (I. A. Genin, ed.), *Narodnoe Khozyaistvo S.S.S.R. Statisticheskiy Sbornik*, (Moskva: Gosstatizdat, 1956), p. 17.

48. I. A. Kulev, *op. cit.*, p. 8.

49. Computed from data in: I. S. Vasilkov, *Razvitie Elektroenergetiki S.S.S.R. Za 40 Let* (Moskva: Gosenergoizdat, 1957), p. 80.

50. I. T. Svets', "Energetichni Problemi Radyanskoi Ukraine Na Novu Pyatilitku," *Visti Akademii Nauk S.S.S.R.*, No. 8, 126 (1946), p. 34 and I. T. Svets *Visti Akademii Nauk S.S.S.R.*, Vol. 25, No. 12, 217 (Dec. 1954), p. 38.

51. I. Grushetskii, "Novom Ugolnom Basseine," *Pravda*, No. 336 (13999), 1 Dec., 1955, p. 2, and M. Odinets, "V Novom Ugolnom Basseine," *Pravda*, No. 21, Oct. 1955, p. 3.

52. P. N. Stepanov, *Ural* (Moskva: Geografgiz, 1957), map p. 62.

53. E. I. Zubtsov, "Formation of the Eastern Fergana Coal Basin," *Materials on the Geology and Mineral Resources of Middle Asia, Leningrad All-Union Geological Institute, No. 10* (1956), p. 32. Prospective development of this basin is excellent according to this author. He also points out that continuous surveying will reveal greater reserves.

54. *Pravda*, No. 31 (14425), 31 January 1958, p. 1. The cost of construction of this plant, begun in 1952 and completed in 1957, was criticized by F. Kleimenov, Chief Engineer, Ministry of the Coal Industry, in *Pravda*, No. 265 (13928), 21 September 1956, p. 2. The annual capacity is 2.5 million cubic meters of gas, the equivalent of 700 thousands tons of coal. *Pravda*, No. 232 (13895), 16 August 1956, p. 1.

55. Averitt and Berryhill, *op. cit.*, p. 26, and Coal Appendix – Table II.

56. A. Zademidko, *op. cit.*, p. 2.

57. Pyotyr Antropov, *Mineral Wealth of the U.S.S.R.* (Moscow: Foreign Languages Publishing House, 1956), p. 69.

58. A. Tolpyshev, *Pravda*, No. 158 (13821), 6 July 1956, p. 3.

59. G. N. Cherdantseva, ed., *Ekonomicheskaya Geografiya S.S.S.R. - R.S.F.S.R.* (Moskva: Gosudstvennoe Uchebno-Pedagogicheskoe Izdatelstvo, 1956), p. 467.

Chapter II

PRODUCTION AND CONSUMPTION OF SOVIET COAL: THE DISTRIBUTION OF PRODUCTION IN RELATION TO POTENTIAL

Coal is the most important single energy resource in the Soviet Union. When all nonrenewable energy resources are converted to a single energy standard, it alone accounts for 94.7 per cent of the total energy potential. [1] Lenin emphasized its importance when he wrote: "... coal is the actual bread of industry; without this bread industry cannot function..." [2] Practically every Soviet book and article on the coal industry is prefaced with this statement. While this illustrates the required familiarity of Soviet writers with the master's works, its current use is indicative of the continuing significance of coal to the Soviet economy.

For over a century, there has been a steady increase in coal production within the territory that now comprises the Soviet Union. During most of this time, the increase in production has been dependent upon mining activities in the Donets Basin. Only since the assumption of power by the Soviets has there been evidence of the decreasing importance of the Donets. The Soviet period has been characterized by increases in the proportionate share of lower grade coals and mining by the open pit method; there has been some migration of mining activity to Eastern regions, but this has been paralleled by a continuing reliance on European sections as the main source of fuel. The consumption of coal remains centered in European Russia, and the bulk of the coal mined in the Eastern areas is shipped great distances to satisfy the demands of consumers in established European industrial centers. Production and consumption of coal in the Soviet Union are currently concentrated in areas of low reserves and a low kilowatt hour energy potential, revealing the nonrational regional distribution of the industry.

THE STRUCTURE OF COAL CONSUMPTION IN THE SOVIET ECONOMY

The following listing illustrates that three branches of the Soviet economy: railroad transportation, the ferrous metals industry, and power generation, consume over 65 per cent of all coal produced.

Consuming Branch of the Economy	Per Cent of Total Production by Consumer 1954[a]	1958[c]	1965[c] (Plan)
Railroad Transport	22.96	17.3	1.2
Ferrous Metallurgy			
(Power and Coke-Chemical)	21.38	--	--
Incl. Coke burned	(15.00)	15.5	21.1
Electrostations	18.53 [b]	32.8	41.1
Machine Building	6.02	--	--
Coal Industry	4.69	--	--
Domestic Uses	--	13.1	18.3
Other Industries and Agriculture	--	21.3	18.3
All Other Branches of the Economy	26.42	--	--
Total	100.0	100.0[d]	100.0

[a]Adapted from G. D. Bakulev, op. cit., p. 39, G. F. Mikheev, *Ekonomika Ugolnoy Promyshlennosti* (Kiev: Gostekhizdat, Uk. S.S.R., 1957), p. 8.

[b]Not included in this figure is the amount of coal used in 791 coal based thermal stations on state and collective farms.

[c]Adapted from D. I. Maslakov, *Toplivnyy Balans S.S.S.R.* (Moskva: Gosplanizdat, 1960), p. 36.

[d]Figures are for 458 mil. tons of the 495.8 mil. tons produced in 1958. Not included would be coal exported and coal reserved for future use.

Although the railroads have ranked first as a consumer of coal in the Soviet Union, a drastic shift downward in this position is planned by 1965. In 1956, 82.6 per cent of the freight turnover on railroads was by steam haulage. [3] While firewood supplied some of the energy for these steam driven locomotives, coal was the major fuel on railroads. Planned electrification of railroads and increased use of diesel engines will result in a reduction of the use of coal by railroads from 79 million tons in 1958 to approximately 7 million tons in 1965. In traveling from Sochi, on the Black Sea Coast of the Caucasus, to Kiev, in the Ukraine, in the summer of 1960, the author noted that the entire route was in some stage of electrification.

With the expansion of economic activity in the Soviet Union, the demands for ferrous metallurgical products will increase, thus creating larger coal requirements. These demands can be satisfied only by an increase in production of high quality coals.

Within the framework of power generation, coal is the leader, producing 64.1 per cent of all electricity generated in 1955. This constitutes over 109×10^9 of the total 170×10^9 kilowatt hours of electricity generated in that year, or approximately 543 out of a total of 850 kilowatt hours produced per capita. [4] The same year the installed capacity of coal based thermoelectric stations was 53.0 per cent of the total 37.2×10^6 kilowatt capacity.

The coal industry is one of the major consumers of its own product. In 1954, the industry consumed 4.7 per cent, or 15 million tons of the total coal produced. This includes coal used to produce power for the mining of other coal. Statistics on the amount of coal used by the chemical industries would be significant, but are unobtainable. Soviet publications imply that there has been an impressive increase in the use of coal by these industries beginning in the spring of 1958. Growth in the major branches of the Soviet economy, and thus a greater dependence on coal as a source of power, is reflected in the growth of the coal industry.

TRENDS IN SOVIET COAL PRODUCTION

Increase in Total Production.—For over 100 years Russian and Soviet coal production has exhibited a constant increase (see Fig. 4). Until the advent of World War II, the Donets Basin alone accounted for most of this increase. From 1855 to 1938, the Donets seldom failed to produce at least two-thirds of the total coal, and in many years its production was well over 80 per cent of the total. It was not until the introduction of the Soviet Five-Year Plans that any decrease in the dependence on this basin became evident. With the invasion of the Ukraine in World War II, and the scorched earth policy of retreating Soviet armies, production virtually ceased. Reconstruction after the war has been phenomenally rapid. Since 1950, the basin has produced over a third of the total coal annually. Although the Donets is now responsible for a smaller percentage of the total coal production, in actual tonnage it produces much more than it did in previous decades, as shown by the listing at the top of page 44.

Since 1913, the year in which most Soviet statistical compilations begin, the production of all types of coal has increased by 1,242.0 per cent (see Fig. 5). A study of Figs. 4 and 5 reveals that two serious reversals in the progressive increase in production have occurred: the first during the Civil War period in the early 1920's, and the second during World War II when the German armies invaded the Ukraine. Both reversals are indicative of the prominence of mining activities in the Donets Basin.

Among the mining regions which have pre-empted the singularly important position of the Donets Basin and contributed to the pro-

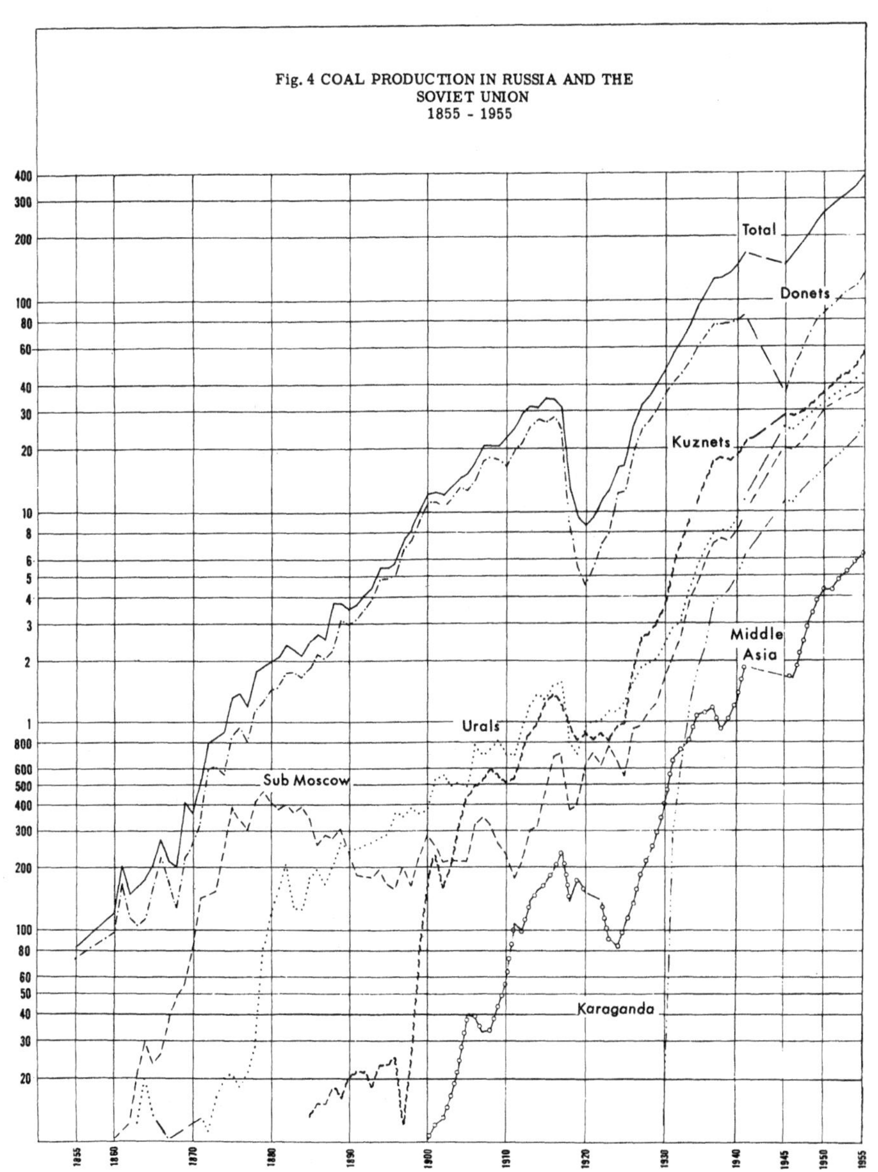
Fig. 4 COAL PRODUCTION IN RUSSIA AND THE SOVIET UNION 1855 - 1955
400
300
200
100
80
60
40
30
20
10
8
6
5
4
3
2
1
800
600
500
400
300
200
100
80
60
50
40
30
20
Total
Donets
Kuznets
Middle Asia
Urals
Sub Moscow
Karaganda
1855
1860
1870
1880
1890
1900
1910
1920
1930
1940
1945
1950
1955

Selected Years	Per Cent of the Total Production	Actual Tons Produced 10³[a]
1855	89	75.7
1875	65	843.0
1895	90	4,886.0
1915	85	26,645.0
1935	63	69,315.0
1938	61	78,330.0
1940	56	85,509.0
1945	26	36,934.0
1950	36	89,679.0
1955	36	135,334.0
1958[b]	36.6	181,662.0

[a]G. D. Bakulev, *Razvitie Ugolnoe Promyshlennost Donetskogo Basseina* (Moskva: Akademiya Nauk SSSR—Institut Ekonomiki, 1955), p. 668, and Ts.S. U.S.S.R. (A. G. Perdukhin, ed.) *op. cit.*, pp. 32-33.

[b]Ts. S.U. Pri Sovete Ministrov S.S.S.R., *Narodnoe Khozyaistvo SSSR v 1958 Gody Statisticheskiy Ezhegodnik* (Moskva: Gosstatizdat, 1959), p. 203.

gressive increase in production are the Kuznetsk and Karaganda Basin, the Urals and Middle Asian regions, and the Moscow Basin.

Trends in Production by Type of Coal.—Parallel to the increase in over-all coal production has been the Soviet emphasis on brown coal mining operations. Brown coal production rose during this period by 9,730.2 per cent. Anthracite coal, which decreases most sharply with the interruption of mining in the Donets, has increased in production by 1,171.6 per cent. Soviet hard coal, roughly equivalent to our bituminous grades, has had the least increase in production. Its growth has been but 847.4 per cent. World War II produced no adverse effect on the continuing growth in production of brown coal. Even though the Moscow basin suffered from invasion, growth persisted because the Soviets accelerated brown coal mining in the Urals, Karaganda, and Eastern Basins (see Coal Appendix - Table IV).

The following listing presents, by types of coal, changes in the nature of coal production since 1913.

Type of Coal Produced	As a Per Cent of Total Production 1913	As a Per Cent of Total Production 1955	Increase in Tons Produced 1913-1955 Per Cent
Brown	3.9	29.9	9,730.2
Hard (Bituminous)	79.7	54.8	847.4
Anthracite	16.4	15.3	1,171.6

The trend toward a greater production of brown coal is the result of the coal industry's desire to utilize local fuels. Brown coal is mined and used locally in the Karaganda Basin, whereas the high en-

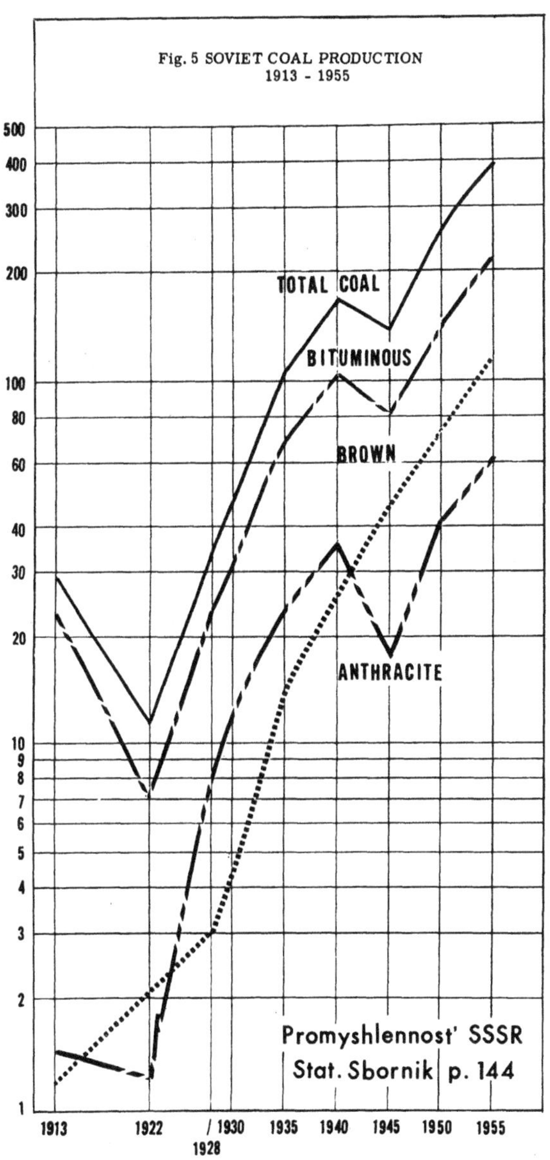

Fig. 5 SOVIET COAL PRODUCTION 1913 - 1955

ergy coal of this basin is shipped elsewhere for consumption. The same situation exists in the Urals.

Growth in the Development of Open Pit Mining.—The amount of coal mined annually by the open pit method increased from 0.6 per cent of all coal mined in 1913 to 19.9 per cent in 1958. Most of this increase occurred in Middle Asian, Siberian and Far Eastern regions (see Coal Appendix - Table IV). Lower production costs and greater labor productivity compelled the Soviets into wider development of open pit operations. N. G. Feitelman in writing on the cost of coal and ways of lowering it, said, [5]

> The productivity of labor in coal cuts is four to six times more than in mines, and the cost of a ton of coal, produced by the open pit method, constitutes only about 30% of the cost of a ton of coal, produced by the underground method.

In addition to increasing labor productivity and lowering costs, the establishment of open pit mines permits the Soviets to accelerate their coal production and at the same time decrease the amount of material needed in mine construction. Professor N. Melnikov summed up all three of these factors when he wrote:

> The productivity of labor in coal cuts is four to five times higher than in mines, and the cost of a ton of coal is three times lower.
>
> The construction of coal cuts is progressing two to three times faster than the construction of mines, and the cost is approximately one half to two times as cheap. In the underground production of each thousand tons of coal, there are expended 35 to 40 cubic meters of wood cripples. In open pits the utilization of wood decreases by ten times.
>
> Reserves of coal deposits in the Kuzbas will permit the construction in the next year of a greater number of large and average size coal cuts with a total annual production of several tens of million tons of coal. When compared to mine construction it will be possible to reduce the capital expenditure by three million rubles, obtaining annually an economy of lower cost in the production of coal by 500 million rubles and decreasing the utilization of wood by one million cubic meters a year. [6]

Productivity increases, with lowered cost, are especially welcome by the coal industry which has long operated at a loss within the framework of Soviet financing. [7] Open pit mining contributes substantially toward lowering costs for the entire coal industry and directing it toward solvency. The following listing adapted from Feitelman reveals that, with each 1.0 per cent increase in open pit mining, there is approximately an 0.8 per cent decrease in cost for the entire coal industry. [8]

Year	Per Cent of Total Production by the Open Pit Method	Per Cent by which Total Costs Were Lowered in the Coal Industry
1949	10.5	7.2
1950	10.9	8.0
1951	11.2	8.1
1952	12.1	8.7
1953	13.7	9.6
1954	15.6	11.1
1955	17.2	12.9

Trends in mining are interrelated. The recent rise in open pit mining coincides with the growth in the production of brown coal; a greater proportion of the total production now consists of low energy and low quality coals. This type of coal is limited in use and relegated to the category of a local fuel. Because most of the open pit mining of this coal is in Eastern regions, it can be assumed that industries, capable of utilizing low grade fuels in quantity, will move in this direction; furthermore, regional patterns of coal production will also undergo changes.

THE DISTRIBUTION OF COAL PRODUCTION AND CONSUMPTION IN RELATION TO ENERGY POTENTIAL AND QUALITY, 1955

In 1955, both Soviet coal production and consumption were centered in European Russia, including the Urals and Caucasus. The vast Soviet coal energy potential, however, is located east of the Urals. This antipodean situation has been historically as well as recently true. I. A. Kulev sums up the condition as it pertains to all energy resources:

> By virtue of a series of historical conditions the basic portion of the productive industries of the U.S.S.R. are concentrated in the European part of the country. Here up to 80 per cent of the fuel and electro-energy are consumed, at the same time 88 per cent of all energy resources are concentrated to the East of the Urals. [9]

The geographic pattern of current production and an evaluation of its evolution is essential to an understanding of Soviet mining accomplishments.

THE REGIONAL PATTERN OF COAL PRODUCTION, 1955

Coal production exemplifies the preceding statement of I. A. Kulev. In 1955, over 65.3 per cent of the production was centered in European Russia, the Urals, and Caucasia, as the listing below reveals. Similar production patterns existed in 1950 and 1958.

Traditionally Established Regions	Per Cent of Geological Reserves[a]	Per Cent of Kwt. Hr. Energy Potential Mineable Reserves[b]	Per Cent of Total Production 1955[c]	Per Cent of Total Production 1958 Preliminary Report[f]
European Russia	7.50	8.30	52.2	52.6
Urals	0.08	0.06	12.4	12.3
Caucasus	0.02	0.02	0.7	0.6
Kazakhstan	1.60	1.49	7.1[d]	6.1[d]
Middle Asia	0.40	0.50	1.7	1.6
South Siberian Belt (Kuznetsk only)	10.50	13.60	15.0	15.2
Eastern Siberia Includes Transbaikal and South Siberian Belt (excluding Kuznets)	79.30	75.50	6.2	7.6
Far East	0.60	0.53	4.3	4.0
	100.00	100.00	99.6[e]	100.0

[a]Computed from Coal Appendix – Table I.

[b]Computed from Coal Appendix – Table II.

[c]Adapted and computed from A. G. Peredukhin, *op. cit.*, p. 37 and Coal Appendix - Table IV.

[d]Includes Karaganda and Ekibastuz only.

[e]Original data in A. G. Peredukhin total 97.6. Two per cent was added for Ukrainian brown coal.

[f]Computed from A. F. Zasyadko, *Toplivno – Energeticheskaya Promyshlennost S.S.S.R.* (Moskva: Gosplanizdat, 1959), pp. 95-96.

Collectively, these European regions possess but 8.32 per cent of the geological reserves, and only 8.38 per cent of the kilowatt hour energy potential. The vast territories to the east of the Urals have over 91.0 per cent of the total coal reserves and energy potential, but were the site of only 34.3 per cent of the coal mined in 1955. Viewed in another manner, a total of 93.8 per cent of the coal production occurs in regions which have less than 25.0 per cent of the geological reserves. In other words, only 6.2 per cent of the coal is produced in regions which contain 80.0 per cent of the geological reserves and over 75.0 per cent of the energy potential. While these facts reveal the gap that currently separates theory from reality in the Soviet coal industry, a complete geographic realization of their significance demands a closer look at intraregional production.

European Russia, excluding the Urals and Caucasus, was responsible for 52.2 per cent of the Soviet coal mined in 1955 (see Fig. 6). Three major coal basins and several small brown coal deposits account for all of the European production, which was centered in the southern section of the European region, primarily the Ukraine. Of the 197.4 × 10^6 tons of coal mined, 75.3 per cent was hard coal and 24.7 per cent was brown coal. All but several million tons of brown coal in the Ukraine were obtained by underground mining (see Coal Appendix - Table IV).

The Donets Basin.—Foremost among the European basins, and for that matter the U.S.S.R., is the Donets Basin. The following listing attests to its dominant position:

Basins and Deposits	Tons Produced 1955 10^3	Per Cent of European Production	Per Cent of U.S.S.R. Production
Donets	135,334.1	68.5	36.0
Moscow	39,301.5	19.9	10.4
Pechora	14,153.4	7.3	3.8
Dneper	8,694.4	4.3	2.0
Total	197,483.4	100.0	52.2

Adapted and computed from Coal Appendix – Table IV.

Increases will inevitably take place in the production of coal in the eastern regions, and the relative position of the Donets, as a major producer, will decline. As previously outlined, this decline is already in process, but it does not diminish the absolute importance of the Donets in the economy of European Russia. Mr. Kirichenko, the foremost Communist official in the Ukraine, expressed this point of view when he said: [10]

> In the economy of the country . . . the Donbas has an especially prominent role. The Donbas is the principal fuel base of the national economy of the European part of the U.S.S.R. In the present year the Ukrainian Donbas will produce 30 per cent of all coal in the country, and over 54 per cent of all coking coal.

The singularly important position of the Donets is derived from its geographical location, in close proximity to iron ores and established industrial centers, and in one of the most densely populated regions of the Soviet Union. Overshadowing this importance are the unyielding facts that the basin has but 2.8 per cent of the geological reserves and only 3.2 per cent of the kilowatt hour energy potential of the country; yet it contains over half of all the mines of the nation. [11]

The Dneper Basin.—The Dneper brown coal basin of the Ukraine also contributes to the centralization of mining activity in the south-

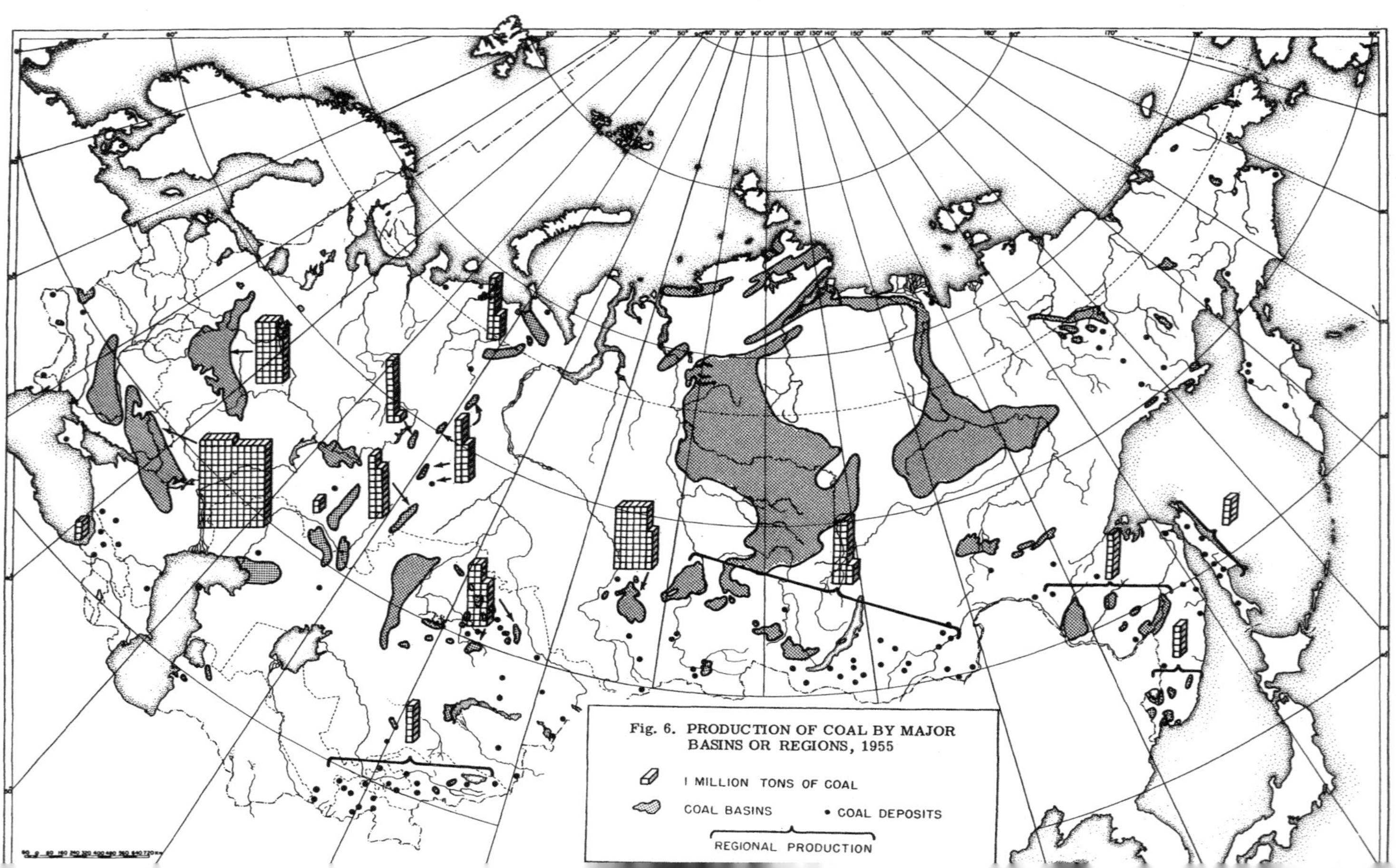

Fig. 6. PRODUCTION OF COAL BY MAJOR BASINS OR REGIONS, 1955

ern section of European Russia. Local fuels constitute 4.3 per cent of the European production; when added to the production of the Donets Basin, the southern section then produces nearly 73 per cent of the total coal tonnage in European Russia. Mining activity has been spasmodic in the basin since its discovery by Professor Levanov in the late 1790's. Industrial production began in 1861 in the Zhurav deposit, and 1862 in the Ekaterinopol deposit. [12] Production ceased altogether in 1903 because of excessive water seepage due to hydrogeological conditions in the mines, and the increasing competition of Donets coal. Some production occurred during World War I and in the 1930's. Extensive industrial production did not take place until after the introduction of large scale open pit methods in 1951. These methods were proposed at a conference held for the development of the basin at Kiev in 1946. [13]

Western Ukraine Brown Coal.—Not all of the brown coal produced in the Ukraine is mined in the Dneper Basin. Approximately 981 thousand tons of brown coal are mined in the Transcarpathian, Precarpathian, and Dneister coal areas of Western Ukraine. [14] Data on the coal mined in the Western Ukraine are not included in the current summaries of coal production by the Ministry of Coal or in the Statistical Handbooks on the economy of the entire U.S.S.R. This coal is produced for local use in small quantities, and it is probably used in power stations on collective farms.

The Moscow Basin.—The Moscow Basin is second in importance as a producer of coal in European Russia, and third in magnitude of coal produced in the entire Soviet Union. Discovered in 1722 by the peasant Ivan Palitsinym, the basin has mined coal industrially in underground workings since 1855. [15] During mining and transportation, this coal crumbles greatly and it has a tendency toward spontaneous combustion when stored for long periods. [16] Low in quality and energy content, these brown coals are used as local fuels. Hydrogeological conditions of water flowing into the mines at a rate of 50 to 300 cubic meters an hour create mining difficulties. [17] The low quality and energy content of the coal, as well as the mining difficulties, do not make it feasible to mine seams under 1.5 meters thick. Fortunately for the Soviets, 95.0 per cent of the layers are more than 1.5 meters in thickness.

Pechora Basin.—Pechora, located in the extreme north, mined approximately 7.0 per cent of the European coal in 1955, or nearly 4.0 per cent of the total Soviet production. With over 53 per cent of the kilowatt hour energy potential of European Russia, and approximately 4.0 per cent of the Soviet Union's geological reserves, this basin has remained relatively unexploited. Two reasons for the delay in development are presented by V. Vityazeva, who states:

> It is impossible, however, to deny a well known obstacle to the development of the coal industry in the comparatively cold Pechora Basin with its fairly severe climate.
>
> .. The former ministry of the coal industry of the U.S.S.R. was erroneously occupied with a somewhat narrow stay-at-home attitude, which did not provide for the proper development of the Pechora Basin, it missed a great opportunity for increasing the output of high quality coals.

Severe climate, coupled with implied bureaucratic bungling, are difficult obstacles to overcome. Nevertheless, the high quality and energy content of the coal should be a stimulus toward accelerating production.

At the present time, the bulk of Pechora coal is utilized in the North and Northwestern sections of the country. Vityazeva, however, advocates its use in the Urals as well, and calls for the completion of the "Pechora-Ural Meridional Railroad." Another strong proponent of this plan is the discoverer of the coking coals in the basin, Dr. A. A. Chernov, of the Komi branch of the Academy of Sciences of the U.S.S.R. He writes that: [19]

> Pechora coals are able to satisfy not only the consumers of the European North, but also industrial enterprises of the Northern Urals and even the Central Urals, where we have a trying need for coals, especially coking. We must rise to the next problem concerning the construction of a railroad to the south from the Pechora Basin along the Northern Urals.

A map showing the proposed route of this railroad appeared in the publication: *Razvitie Proizvoditelnykh Sil Vostochnoy Sibiri, Chernaya Metallurgiya,* (Moskva: Akademii Nauk, 1960) inserted pp. 4-5. From Polunochnoe this line is shown extending north to a point west of Salekhard. A second proposed line is depicted from Polunochnoe northeast to Nizhne Narykary on the Ob River. If completed, either line could deliver Pechora coals to the Urals.

The Urals.—The Urals produced 46.8 million tons of coal in 1955, or 12.4 per cent of the nation's total. In 1955, brown coal accounted for 74.3 per cent, while hard coal, including 0.8 per cent anthracite, made up the remaining 25.7 per cent. In the Urals, 50.5 per cent of the coal mined came from open pits. Even though the Urals ranks fairly high as a coal producer, its geological reserves are relatively insignificant, comprising but 0.08 per cent of the Soviet total. The kilowatt hour energy potential of the reserves is but 0.06 per cent of the Union's total. This figure is low because of the high percentage of brown coal.

Production is scattered throughout many small deposits, but the administration of this production is centralized in four mining combinats. Hard coal is obtained almost exclusively from the Kizel Basin, under the direction of the Molotov Coal Combinat, as indicated in the following listing:

Mining Combinats on 1/1/58	Tons Produced 1955 10^3	Per Cent of Urals Production	Per Cent of U.S.S.R. Production
Molotov (Kizel)	11,020.0	23.6	2.92
Chelyabinsk	17,639.6	37.7	4.68
Sverdlovsk	16,361.4	34.9	4.33
Bashkir	1,834.7	3.8	0.47
Total	46,856.6	100.0	12.40

A small amount of the hard coal, including all of the anthracite, is produced by the Sverdlovsk Coal Combinat. Brown coal exploitation in the Bashkir Region by the Bashkir Coal Combinat is a late post World War II development.

P. N. Stepanov claims that the Urals have over 1,000 different minerals located in over 12,000 deposits. [20] Brown coal, burned in thermoelectric stations, supplied much of the energy used in converting these minerals to manufactured goods. Industry consumed over 90 per cent of the electricity produced in the Urals in 1956. [21] This is well above the Soviet national average for industry. Makushkina, and others, justify this with the unequivocable claim that: "The Urals is the largest industrial region of the Soviet Union." [22]

Caucasia.—Caucasian coal production is under the administration of the Georgian Coal Combinat. This includes the two hard coal basins of Tkibuli and Tkvarcheli, and the brown coal deposit of Akhaltsikh. In 1955, this region mined 0.7 per cent, or 2.7×10^6 tons of the Soviet coal. Hard coal, coked for use in local metallurgical plants, is the major fuel mined. Brown coal was not mined on an industrial basis before World War II; however, by 1955 it was 5.4 per cent of the combinat's output. Geological reserves and kilowatt hour energy potentials do not exceed 0.02 per cent of the nation's total.

Kazakh Coal Deposits.—Kazakhstan mined 7.1 per cent of the Soviet coal in 1955. The 26.8×10^6 tons of coal represented by this per cent are from the Karaganda and Ekibastuz deposits. Approximately 23 per cent was brown coal. Open pit mining, confined exclusively to the Ekibastuz deposit, amounted to 31.1 per cent of all production. Half of the underground extraction in the Karaganda Basin came from two coal seams; the Feliks, which is 4.5 meters thick; and the Verkhnyaya Marianna, which is 8 meters thick. [23] Kazakhstan possesses 1.6

per cent of the total energy potential. Quality of Karaganda and Ekibastuz coals is excellent.

Coal was also mined in Kazakhstan in 1955 by organizations other than the Karaganda Coal Combinat as listed below:

Oblast Producing[a]	Tons Produced 10^3	Type of Coal Produced[b]
Akmolinsk	126.1	Brown
Aktyubinsk	90.3	Hard
Gurev	4.8	Brown
Dzhambul	17.9	Brown
South Kazakhstan	693.3	Brown

[a]A. Z. Gravarnik, *National Economy of the Kazakh S.S.R. Statistical Handbook* (Alma-Ata: Kazgosizdat, 1957), p. 36.

[b]Ya. A. Magolin, ed., *The Kazakh S.S.R., Economic-Geographic Characteristics* (Moskva: Geografgiz, 1957), map p. 19.

With the exception of the South Kazakhstan Oblast, this production is not included in the compendium on coal production in Coal Appendix - Table IV. The South Kazakhstan Oblast contains the Lenger and Karatay deposits which are under the jurisdiction of the Middle Asia Coal Combinat, and their production is reported by that combinat. Admittedly insignificant in amount, this production is probably part of the 0.4 per cent unaccounted for in the listing on page 48. In all probability, the use of this coal was restricted to the area of production.

Middle Asian Production.—Middle Asian coal extraction is scattered throughout numerous small deposits. The region's total production of 6.3×10^6 tons, or 1.7 per cent of the Soviet total, is less than brown coal production in the Ukraine. In 1955, hard coal accounted for 15.5 per cent of the production and brown coal 84.5 per cent. Nearly 34 per cent of the extraction occurred in open pit mines.

Statistics on Soviet coal production are reported by combinat: when the combinat includes only one basin, production can be pinpointed. The Republics of Middle Asia and the South Kazakh Oblast are under the administration of a single combinat. From a geographer's viewpoint, this leaves much to be desired. It has been possible, however, for the author to compute this production on a republic basis for 1955, as indicated in the table on the facing page.

With the exclusion of Turkmen S.S.R., all Middle Asian Republics produce some coal. Quality of coal varies with the individual deposits. The energy potential of the entire region is but 0.50 per cent of the Soviet total.

The Kuznetsk Basin.—The Kuznetsk Basin is the second largest coal producing region in the U.S.S.R. In 1955, the 56.5 million tons of

Republic or Oblast	Per Cent of Middle Asian Production[a]	Tons Produced[b]
Uzbek	41	2,596,571
Kirgiz	40	2,533,240
Tadzhik	8	506,648
South Kazakh Oblast	11	696,641[c]
Total	100	6,333,100

[a]B. M. Fofanov, "Development of the Coal Industry in Middle Asia," *Nauchn Zap. Tashkentsk Finas – Ekonomika Institut*, 1958, Vyp. 10 (1958), 105-113.

[b]Computed from percentages given in column one above, and Coal Appendix Table IV.

[c]A discrepancy of 3.3×10^3 tons for the South Kazakh Oblast exists between this listing and the preceding one. The writer is unable to determine which source contains the error.

coal mined here comprised 15.0 per cent of the Soviet coal. All of the production is hard coal, much being suitable for coking purposes. In 1956, the Kuzbas supplied 29 per cent of the coking coal in the country, [24] or 23.8 million tons. Production goals for the basin in 1960 are 88 million tons of coal, including 32 million tons of coking coal. [25]

Most of the mining takes place in underground shafts. In 1955, exactly 90 per cent of the coal was produced in this manner. Exceptionally thick seams in steeply pitched synclines, particularly in the southern section of the basin, often require especially adapted mining techniques and machinery. This is true in the mines of the Prokopev-Kiselev region, where the total thickness of the coal layers ranges from 16 to 76 meters. [26] In the northern section of the basin, the seams are less thick and almost horizontal; many are near the surface and can be mined from open pits. Open pit mining, which comprised 10 per cent of production in 1955, is to be expanded. The quality and amount of coals which can be mined in this fashion are excellent, as S. Kuryshkin states:

> Reserves of coal, suitable for excavation by the open pit method, in the Kuznetsk Basin, are very great. Here it is even possible to produce coking coals by the open pit method. [27]

Soviet planners expect to produce 20×10^6 tons of coal from open pits in the Kuznetsk Basin by 1960. This would be 22.7 per cent of the basin's total coal production for that year. [28]

East Siberian Coal Production.—Eastern Siberia produced 6.2 per cent of the Soviet coal in 1955. The East Siberian Coal Combinat responsible for this production had jurisdiction over all basins and deposits west of the Kuznetsk Basin through the Chita Oblast, and south of the Tunguska Basin to the border of Mongolia. Of the 23.1×10^6 tons of coal produced in this region, 70 per cent were hard coal and

30 per cent brown coal. Much of the mining, 47 per cent in 1955, is accomplished in open pit workings.

Because of the Soviet method of reporting production by combinat rather than basin, it has not been possible to depict a detailed distribution pattern within this region for 1955. However, an established production pattern is reflected in partial data for the following year. This data, presented below, is approximate, and does not include production for the Chita Oblast.

Region or Basin	Tons Produced in 1956
Buryat Mongolian A.S.S.R.	870,000[a]
Irkutsk Basin	14,100,000[b]
Kansk-Achinsk Basin	5,000,000[c]
Minusinsk Basin	2,500,000[d]

[a]P. I. Korobov, ed., *The R.S.F.S.R. Over 40 Years, Statistical Handbook* (Moskva: Sovetskaya Rossiya, 1958), p. 165.

[b]S. V. Troyanski, *Gornoe Delo Entskilopedicheskiy Spravochnik Vol. 2* (Moskva: Ugletekhizdat, 1957), p. 386.

[c]*Ibid.*, p. 376.

[d]*Ibid.*, p. 381.

These figures suggest that two-thirds of the production is centered in the Irkutsk Basin. Troyanski states that 93 per cent of the output of this basin is from the Cheremkovo Deposit. Little production takes place in the section East of Lake Baikal.

The Far East.—Far Eastern coal deposits contributed 4.3 per cent of the Soviet production in 1955, or 16.0×10^6 tons. Brown coal accounted for 69.4 per cent of the production and hard coal 30.6 per cent. A majority of the coal mined in this region, 55.6 per cent, is extracted from underground mines. Geographically, three subregions exist in the Far East; each is under the direction of a separate mining combinat (see Coal Appendix - Table V). Production by these subregions for 1955 is outlined below, as computed from Coal Appendix - Tables IV and V:

Combinat Subregion	Tons Produced 1955 10^3	Per Cent of Far Eastern Production
Far East Coal Combinat	6,992.4	43
Sea Litoral Coal Combinat	5,441.8	34
Sakhalin Coal Combinat	3,622.9	23

The Far Eastern Combinat derives most of its coal from open pit mines; its production is practically all brown coal. In 1955, 93.5 per cent of its production was brown coal from open pit mines. Over 59 per cent of the Sea Litoral's production was brown coal. All coal is extracted from underground coal shafts. On Sakhalin Island, 61.6 per cent of the coal extracted was hard coal; much of it was of coking quality. [29] Over 16 per cent of the production was from open pit operations.

Arctic and Subarctic Siberia and the Northeast.—Arctic and Subarctic Siberia and the Northeast possess 55.8 per cent of the nation's geological reserves of coal and 65.5 per cent of the kilowatt hour energy potential. Yet nowhere in this extensive territory are there mining operations significant enough to be included under the Ministry of Coal as a combinat. Troyanski refers to mining in this region as local mining, and cites river and coastal ships as major consumers. The Norilsk mining region, however, is an exception in his generalization. Production data on this vast region do not appear in the standard Soviet statistical compilations, nor are they included in reports available on the coal industry. Sources that do reveal production quote archival data. Production is often given as an approximation, thus negating its effectiveness for inclusion in a compendium of all Union production derived from more definitive sources. The data presented below are submitted with the intention that they be representative of generalized patterns of production in the region.

Region, Basin, or Deposit	Tons Produced 1954[a]	Tons Produced 1956[b]
Tiksi (Saginsk Deposit)	57,000	48,000
Tunguska Basin (Norilsk Region)	n.d.	2,000,000
Lena Basin – including		
Sengar Deposit (Verkhoyansk)	187,000	194,000
Kangalas Deposit (Yakutsk)	60,000	34,000
Dzhebarkik Deposit	108,000	103,000
South Yakutsk Basin	35,000	n.d.
Zyryansk Basin	136,000	100,000
Arkagalin Coal Region	n.d.	600,000
Elgen Coal Region	n.d.	250,000
Bukhty Deposit	n.d.	130,000

[a]V. F. Vasyutin ed., *Problems in the Development of Industry and Transport in the Yakutsk A.S.S.R.* (Moskva: Akademii Nauk S.S.S.R., 1958), p. 134.

[b]S. V. Troyanski, *op. cit.*, p. 397, 373, 405, 402, 442, 443, 444, 445.

Speculation concerning the ultimate use of this coal would be scientifically unsound. A lack of transportation facilities, however, other than river barges, precludes extensive inland distribution of the coal. Norilsk, with the largest production in the region, is also the site of nonferrous metallurgical works. Much of the coal extracted here is destined for these plants.

The Eastward Migration of Production.—Under the Soviet regime, a substantial segment of the mining activity has shifted eastward from the traditional European centers of production (see Table 4).

TABLE 4

SOVIET COAL PRODUCTION BY BASIN OR REGION, SELECTED YEARS 1913-1958

(in per cent)

Region or Basin	1913[a]	1927-28[a]	1932[a]	1937[a]	1940[a]	1945[a]	1950[a]	1955[a]	1958[c]	1965[c] Plan
European Russia										
Donets Basin	86.8	75.9	70.4	60.3	55.8	25.8	36.1	36.0	36.6	36.9
Moscow Basin	1.0	3.3	4.2	6.0	6.5	14.0	12.3	10.4	9.5	5.8
Pechora Basin	---	---	---	0.1	0.2	2.3	3.5	3.8	3.4	3.1
Ukraine Brown Coal	---	---	---	---	---	---	---	2.0[b]	2.6	1.9
Lvov-Volyn	---	---	---	---	---	---	---	---	0.4	1.6
European Total	87.8	79.2	74.2	66.4	62.5	42.1	51.9	52.2	52.5	49.3
Caucasia (Georgian Coal Combinat)	0.2	0.2	0.3	0.3	0.4	0.5	0.7	0.7	0.6	0.6
Urals	4.2	5.9	5.1	6.5	7.6	17.5	12.9	12.4	12.3	9.6
European Total incl. Caucasia and Urals	92.2	85.3	79.6	73.2	70.5	60.1	65.5	65.3	65.4	59.5
Eastern Regions										
Karaganda Basin (All Kazakhstan)	---	---	1.2	3.2	4.1	7.9	6.5	7.1	6.3	8.1
Middle Asia	0.9	0.8	1.2	0.8	1.3	1.2	1.7	1.7	1.6	1.9
Kuznetsk Basin	2.7	7.7	10.9	13.9	13.8	20.3	14.8	15.0	15.2	16.8
East Siberia including Transbaikal	2.9	3.0	3.9	4.6	5.6	5.3	6.0	6.2	7.3	9.1
Far East	1.3	3.2	2.7	3.8	4.3	4.9	4.8	4.3	4.0	4.2
Eastern Total	7.8	14.7	19.9	26.3	29.1	39.6	33.8	34.3	34.4	40.1
Total Soviet Production	100.0	100.0	98.5	99.5	99.6	99.7	99.2	99.6[d]	99.8	99.6

[a]Adapted and computed from A. G. Peredukhin, ed. , *Ugolnaya Promyshlennost S.S.S.R. Statisticheskiy Spravochnik* (Moskva: Ugletekhizdat, 1957), p. 37 and Coal Appendix – Table IV.

[b]Computed from data in Coal Appendix – Table IV.

[c]D. I. Maslakov, *Toplivnyy Balans SSSR* (Moskva: Gosplanizdat, 1960) p. 63.

[d]Totals were computed from the data presented in Peredukhin. Where totals do not equal 100 per cent, the author can only assume that the balance was local production not under the jurisdiction of the Ministry of Coal.

Prior to the revolution, European Russia, including the Urals and the Caucasus, produced 92.2 per cent of the coal, while areas east of the Urals produced only 7.8 per cent. Since that time, the eastern regions have registered a progressive increase in their proportionate share of the coal mined. By 1955, European Russia was producing only 65.3 per cent, while the eastern regions produced more than a third of the nation's coal. Each of the individual basins and regions in the east has shared in the growing per cent of the Union's coal produced there.

Two basins, the Karaganda and the Kuznetsk, however, are responsible for most of the increase in eastern production. Collectively, these basins produced 22.1 per cent of the coal in 1955. Pre-revolutionary production records 2.7 per cent for the Kuznetsk Basin, and production was not begun at Karaganda until after 1928.

Middle Asia's share in the total production increased from less than one per cent to almost two per cent during this period.

The immense territory, from the Kuznetsk Basin east to the Pacific Ocean, has increased in its share of total Soviet production—from 4.2 per cent in 1913, to 10.5 per cent in 1955. According to Communist theory, this vast area with its tremendous resources should be one of the major coal producers in the Union. Yet, by 1955, the region was producing about the same amount of coal as the Moscow brown coal basin.

In over four decades of Soviet power, approximately 27 per cent of the coal production has shifted to the eastern regions. Most of this shift occurred prior to 1940. When considered with the increase in the total coal production, this shift becomes even more important, and lends credibility to the Soviet claim that they are bringing industry closer to the sources of raw material. Credibility can be extended to this claim only as it pertains to the extraction of coal, not its consumption.

The Relative Stability of Consuming Regions and Long Distance Coal Hauls.—Coal consumption in the U.S.S.R. is centered in the Central Region and the Ukraine or European Russia (see Fig. 7). Consumption exceeds local production, and demands the importation of coal from areas in the east. This situation is contrary to the Soviet hope for a closer alliance between production and consumption. Soviet writers have pointed this out in general terms. Vityazeva says, [30]

> However in planning the territorial development of the coal industry, there are still serious deficiencies; as a result of this, the rate of growth of coal mining in the European part of the country lags behind the rate of growth of its consumption. ... Kuznetsk coal is supplied, for example, to such oblasts as Kirov, from a distance of 2790 kilometers, and Gorkie—3189 kilometers.

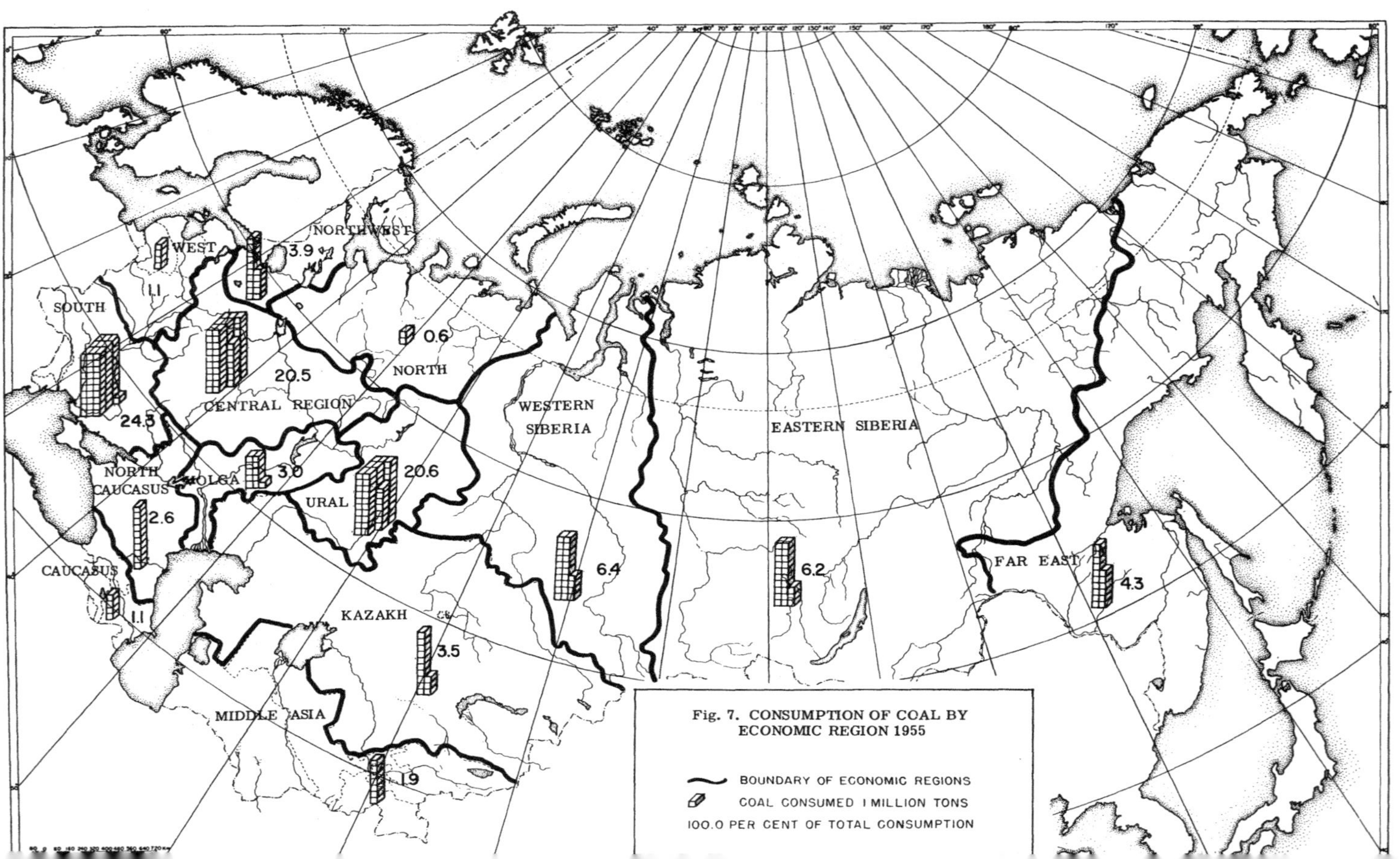

Fig. 7. CONSUMPTION OF COAL BY ECONOMIC REGION 1955

Remarks such as these are not isolated occurrences in Soviet journals. They appear monthly as a continuing testimony to the stability of European Russia as the major coal consuming region. The following listing on European Russia, including the Urals and the Caucasus, indicates that this consumption pattern has not changed in over a decade and a half.

Year	Per Cent of Total Coal Production European Part	Coal Imported from Eastern Russia as a Per Cent of Total Production	Coal Consumed in European Russia as a Per Cent of Total Production
1940	70.5	7.7	78.2
1950	65.5	12.6	78.1
1955	65.3	12.0	77.3
1958[a]	65.5	11.3	76.8

[a]Based on preliminary data.

Coal imports from the eastern basins into European Russia have increased in proportion to the shift in coal production from European Russia to the eastern regions; as a result, the per cent of the total coal consumed in the European area has remained relatively constant. During the period 1940 to 1955 represented by the previous listing, coal production in European Russia decreased by 5.2 per cent, while the importation of coal from eastern regions increased by 4.3 per cent, and the total consumption decreased by only 0.9 per cent. Well beyond the province of this study is the matter of coal exports and imports through European Russia to and from foreign areas. A survey of this type might indicate that European Russia's coal consumption has noticeably increased beyond expectation. The implication of this increase in importation from Eastern basins becomes more important when it is realized that only the high quality, high energy content coals are shipped to European consumers.

Three regions, the South, the Central, and the Urals, consume most of the coal within European Russia (see Table 5). The South is the only self-sufficient region in coal production. Local production accounts for half of the coal consumed in both the Central Region and the Urals. However, this local production is either brown coal or low quality, highly sulfurous hard coal, unsuitable for use in metallurgical plants. Continuance of basic industry in these two regions will depend upon the importation of high energy content coals from distant eastern basins, until the Pechora basin is fully developed and railroad transportation is reoriented.

Regions in European Russia dependent upon imported coal are the Volga, the West, and the Northwest. The Volga region, which con-

TABLE 5

REGIONAL DISTRIBUTION OF SOVIET COAL CONSUMPTION, 1955[a] and 1958

(in metric tons)

Consuming Region and Source of Coal	Tons of Coal Consumed 1955	Per Cent of Consumption 1955	Per Cent of Consumption 1958[c]
The South			
Local Production			
Donets Basin	82,716,201		
Ukrainian Brown Coal	8,694,400		
Total Consumption	91,410,601	24.28	26.0
Central Region			
Local Production			
Moscow Basin	39,301,500		
Imported coal			
Donets Basin	30,057,703		
Kuznetsk Basin	4,415,547		
Kizel Basin	2,413,577		
Karaganda Basin	943,785		
Total Consumption	77,132,112	20.49	16.9
Volga Region			
No local production			
Imported coal			
Donets Basin	4,614,892		
Kuznetsk Basin	3,810,600		
Karaganda Basin	2,858,169		
Kizel Basin	52,900		
Total Consumption	11,336,561	3.01	2.5
The West			
No local production			
Imported coal			
Donets Basin	4,208,890		
Total Consumption	4,208,890	1.12	2.0
The North West			
No local production			
Imported coal			
Donets Basin	2,706,682		
Pechora Basin[b]	11,888,856		
Total Consumption	14,595,538	3.88	3.4
The North			
Local Production	2,264,544		
Total Consumption	2,264,544	0.60	2.2
The Urals			
Local Production	44,081,538		
Imported coal			
Kuznetsk Basin	21,382,331		
Karaganda Basin	11,926,022		
Total Consumption	77,389,891	20.55	19.6
The Northern Caucasus			
Donets coal only	9,608,721		
Total Consumption	9,608,721	2.55	3.0

[a]Computed from data in Coal Appendix – Tables IV and VI.

[b]Computed from data in I. I. Tarabukin, *Narodnoe Khozyaistvo Komi ASSR, Statisticheskiy Spravochnik* (Syktyvkar: Komi Kinizhnoe Izdatelstvo, 1957), pp. 22 and 104.

Consuming Region and Source of Coal	Tons of Coal Consumed 1955	Per Cent of Consumption 1955	Per Cent of Consumption 1958[c]
The Transcaucasus			
Local Production	2,706,200		
Imported coal			
Donets coal	1,353,341		
Total Consumption	4,059,541	1.08	1.2
Middle Asia			
Local Production	6,333,100		
Imported coal			
Kuznetsk Basin	435,335		
Karaganda Basin	329,788		
Total Consumption	7,098,223	1.89	6.1 (Middle Asia and Kazakhstan)
Kazakhstan			
Local Production	10,424,545		
Imported coal			
Kuznetsk Basin	2,681,166		
Total Consumption	13,105,711	3.48	
Western Siberia (Southern Section)			
Local Production	23,717,313		
Imported coal			
Karaganda Basin	329,788		
Total Consumption	24,047,101	6.39	7.5
Eastern Siberia			
Local Production	23,173,300		
Total Consumption	23,173,300	6.15	5.7
Far East			
Local Production	16,057,100		
Total Consumption	16,057,100	4.26	3.9
Coal exported from regions indicated – consuming region unknown.			
Donets Basin	13,533		
Kuznetsk Basin	84,805		
Kizel Basin	308,000		
Total Exported	406,338	0.11	
Coal production and consumption unaccounted for	616,096	0.16	
	376,498,700	100.00	

[c] A. I. Maslakov, *Toplivnyy Balans SSSR* (Moskva: Gosplanizdat, 1960), p. 147.

sumed three per cent of the Soviet coal in 1955, imports from the Donets, Kuznetsk, Karaganda and Kizel Basins. Coal consumed in the West and the Northwest is brought from the Donets and Pechora Basins.

Caucasian regions produced some of their own hard coal in 1955, but still rely on imports from the Donets for fuel.

Middle Asia is virtually self-sufficient in coal production. Kuznetsk and Karaganda high quality coals are imported possibly for use in metallurgical plants, but the amounts are exceedingly small. Development of high quality, high energy content coals in the East Fergana Valley could alleviate the necessity for this importation.

Western Siberia, with its famed Kuznetsk Basin, is the major coal consuming region in the East. In 1955, coal consumed in this region, including infinitesimal imports from Karaganda, constituted 6.39 per cent of the Soviet total. Karaganda imports were used in enterprises which are closer to that basin than they are to the Kuznetsk.

Eastern Siberia and the Soviet Far East produce and consume their own coal. Soviet sources indicate the existence of infinitesimal shipments to the Buryat A.S.S.R. in 1955 and 1956. [31]

Only one area in all of European Russia, the South, has a production sufficient to meet the demand of its consumers. In both the Central and Urals regions, the extraction of local fuel has been developed extensively as an aid in curtailing the importation of coal from distant basins. In spite of the concentrated effort expended on local fuels, these two, and other European regions, are becoming increasingly dependent upon imported eastern coal. Continuing and increasing imports of coal into the European Region imply that long distance coal shipments are an established feature in the Soviet economy.

Long-distance shipments of coal originate in five hard coal basins. These basins have 22 per cent of the kilowatt hour energy potential of the Soviet Union and produced 65 per cent of the coal in 1955. The coal exported and the amount retained for local consumption are depicted, in the following listing, as a percent of the total production for the basins in 1955.

Category	Pechora [a]	Donets [b]	Kizel [b]	Karaganda [b]	Kuznetsk [b]
Consumed Locally	16.10	61.47	74.82	33.05	41.95
Exported to other areas	83.90	38.53	25.18	66.95	58.05
Total	100.00	100.00	100.00	100.00	100.00

[a]Computed from I. I. Tarabukin, ed., *National Economy of the Komi A.S.S.R., Statistical Handbook* (Syktyvkar: Komi Knizhnoe Izdatelstvo, 1957) pp. 22 and 104.
[b]Computed from Coal Appendix — Table VI.

Coal exports from the Donets Basin are shown in Fig. 8. Each major European region, except the Urals, received coal from the Donets. The actual tonnage of coal shipped from the basin in the last decade and a half has increased, even though the per cent of coal retained for local consumption in the South has risen. Decreases in the amount of coal imported from the Donets have occurred in the Northwest. All other regions evidence an increase in coal imported from the Donets during this period. Shipments to the Central Region account for half of the coal exports from the Donets.

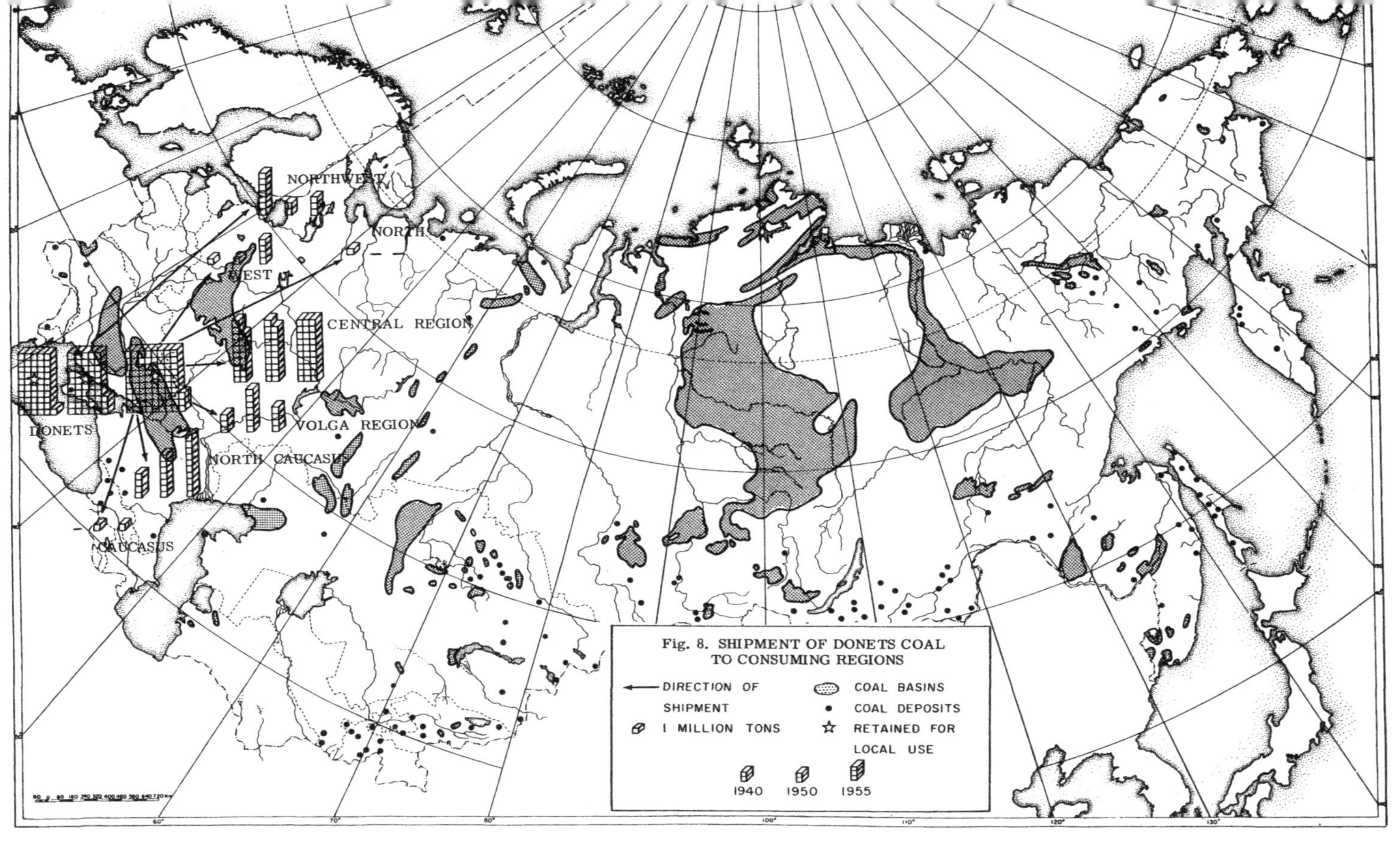

Fig. 8. SHIPMENT OF DONETS COAL TO CONSUMING REGIONS

Shipments of coal from the Pechora Basin are confined to the Northwest. A post World War II development, these shipments now replace Donets coal in this region. Admittedly, the change from Donets to Pechora coal does not reduce the distance from which coal must be hauled, but it has added to the supply of coal available, and permitted the Soviets to establish a metallurgical base utilizing Pechora coal and Karelian ores.

Kizel coal is sent only to the Central and Volga Regions. Shipments from this basin began after World War II. The distances involved in these shipments are not so great as those for other basins, especially eastern ones.

In 1955, the Karaganda Basin sent two-thirds of the coal it produced outside of the Republic's boundaries (see Fig. 9). Shipments to the Urals accounted for 45 per cent of the basin's extraction. Prior to World War II, the Urals received more than half of the Karaganda output; however, as production in the basin grew, tonnages shipped also increased. Nearly 12 million tons of Karaganda coal were shipped to the Urals in 1955. This exceeded the production of the Kizel Basin situated in the Urals. Moreover, shipments from Karaganda to the Volga Region, in the same year, were greater than the entire production of the Transcaucasus Region. Coal exported to the Central Region had reached nearly one million tons at that time. (see Coal Appendix - Table VI).

Shipments from the Kuznetsk Basin are greater, and they travel farther than coal exports from any other deposit (see Fig. 10). Over 58 per cent of the coal produced in the Kuznetsk Basin in 1955 was shipped to other regions. Industry in the Urals received the bulk of this exported coal. In 1955, the Urals received over 21 million tons of Kuznetsk coal, an amount greater than the output of all the Karaganda mines. Shipments to the Central European Region were in excess of four million tons, or more than all the coal consumed in the European West (see Table 15). The Kuznetsk Basin shipped four times as much coal to the Central Region as did the Karaganda Basin; yet Karaganda is closer. Nearly four million tons of coal were shipped from the Kuznetsk to the Volga Region, and almost the same amount was shipped to Kazakhstan and Middle Asia (see Coal Appendix - Table VI).

Past and present Soviet policy has been directed toward establishing industrial regions in parts of the country that contain all, or nearly all, of the necessary raw materials for successful operation of industrial enterprises. [32] One of the principal objectives behind this policy has been the desire to curtail long-distance haulage of freight, especially coal which adds no weight to the finished product. [33]

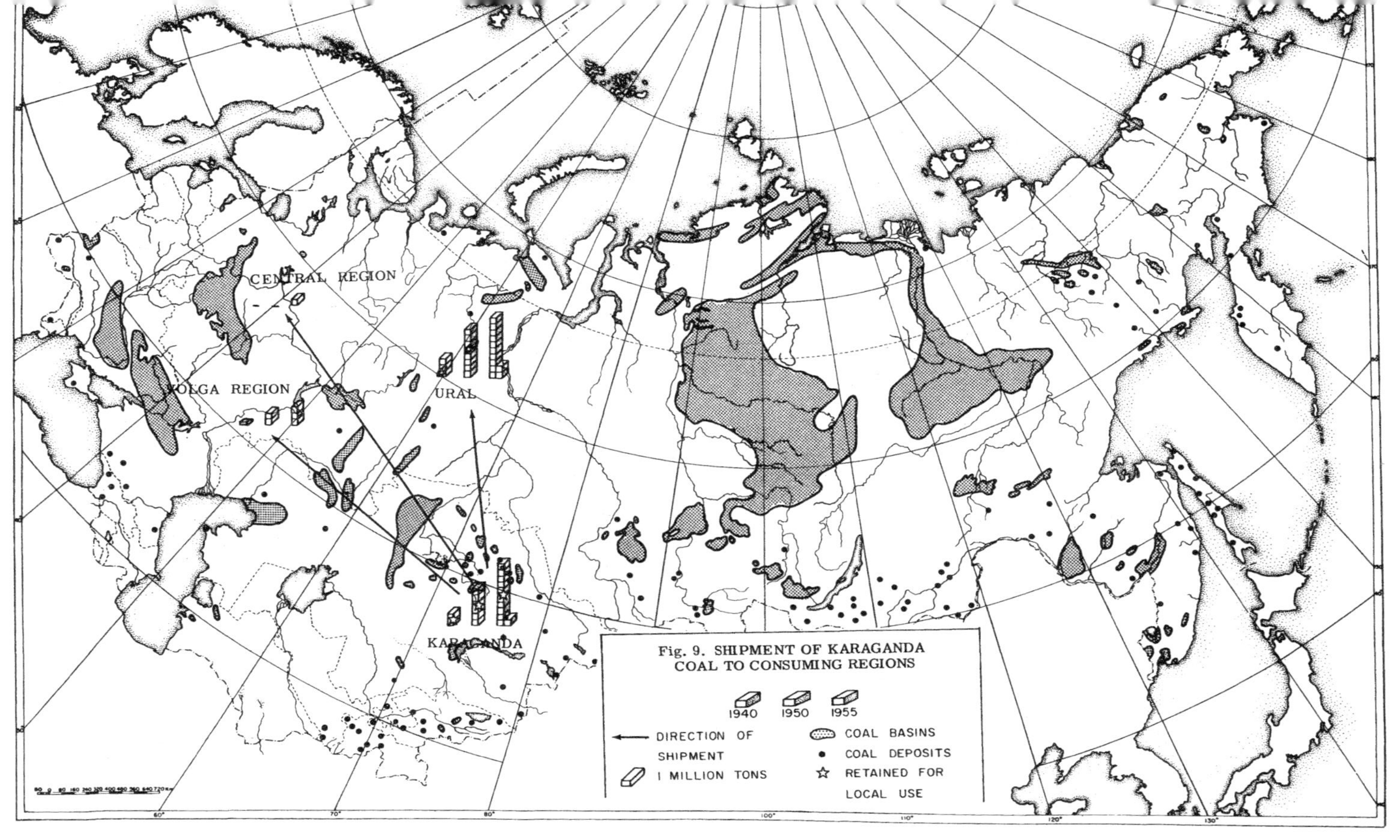

Fig. 9. SHIPMENT OF KARAGANDA COAL TO CONSUMING REGIONS

A. E. Probst, in criticizing the former regime for its tolerance of long-distance coal shipments stated,

> ... the mean rail haul of a coal shipment in 1912 was 529 km. Such hauls are unparalleled in any other country in the world. [34]

When industrial expansion in the Soviet Union created greater demands for fuel, the coal industry in the eastern regions increased production to meet these demands. Coal consumption, however, remained centered in the European part of the country. The per cent of coal consumed in this region has not changed noticeably since before World War II, but the tonnage of coal consumed has increased considerably. In order to satisfy these requirements, larger amounts of coal must be shipped between regions within European Russia and from the distant eastern regions, primarily the Karaganda and Kuznetsk Basins. As a result, inordinately longer freight hauls have evolved. From 1913 to 1956, the coal transported on Soviet railroads increased by 396 million tons, which is equal to an increase in freight turnover of 290 billion ton kilometers (see Table 6). The greatest increase occurred between 1940 and 1956. The average distance for the transportation of a ton of coal increased from 485 kilometers to 729 kilometers during the years 1913 to 1958. This transportation in-

TABLE 6

RAILROAD SHIPMENTS OF COAL AND COKE IN THE U.S.S.R. SELECTED YEARS 1913 - 1956[a]

Years	Freight Turnover (Billions of ton kilometers)	Transported (In millions of tons)	Average Distance One Ton is Transported (kilometers)
1913	13	26	485
1928	19	30	615
1932	38	57	662
1937	83	117	709
1940	107	153	701
1945	99	142	693
1950	178	266	670
1951	188	283	665
1952	201	304	661
1953	212	322	657
1954	235	349	672
1955	267	389	686
1956	303	422	718
1957	--	--	741[b]
1958	--	--	729[b]

[a] Adapted from *Tsentralnoe Statisticheskoe Upravlenie Pri Sovete Ministrov SSSR* (K. G. Ivanova, ed.), *Transport i Svyaz SSSR Statisticheskiy Sbornik* (Moskva: Gasstatizdat, 1957), p. 33.

[b] Adapted from D. I. Maslakov, *Toplivnyy Balans S.S.S.R.*, (Moskva: Gosplanizdat, 1960), p. 150.

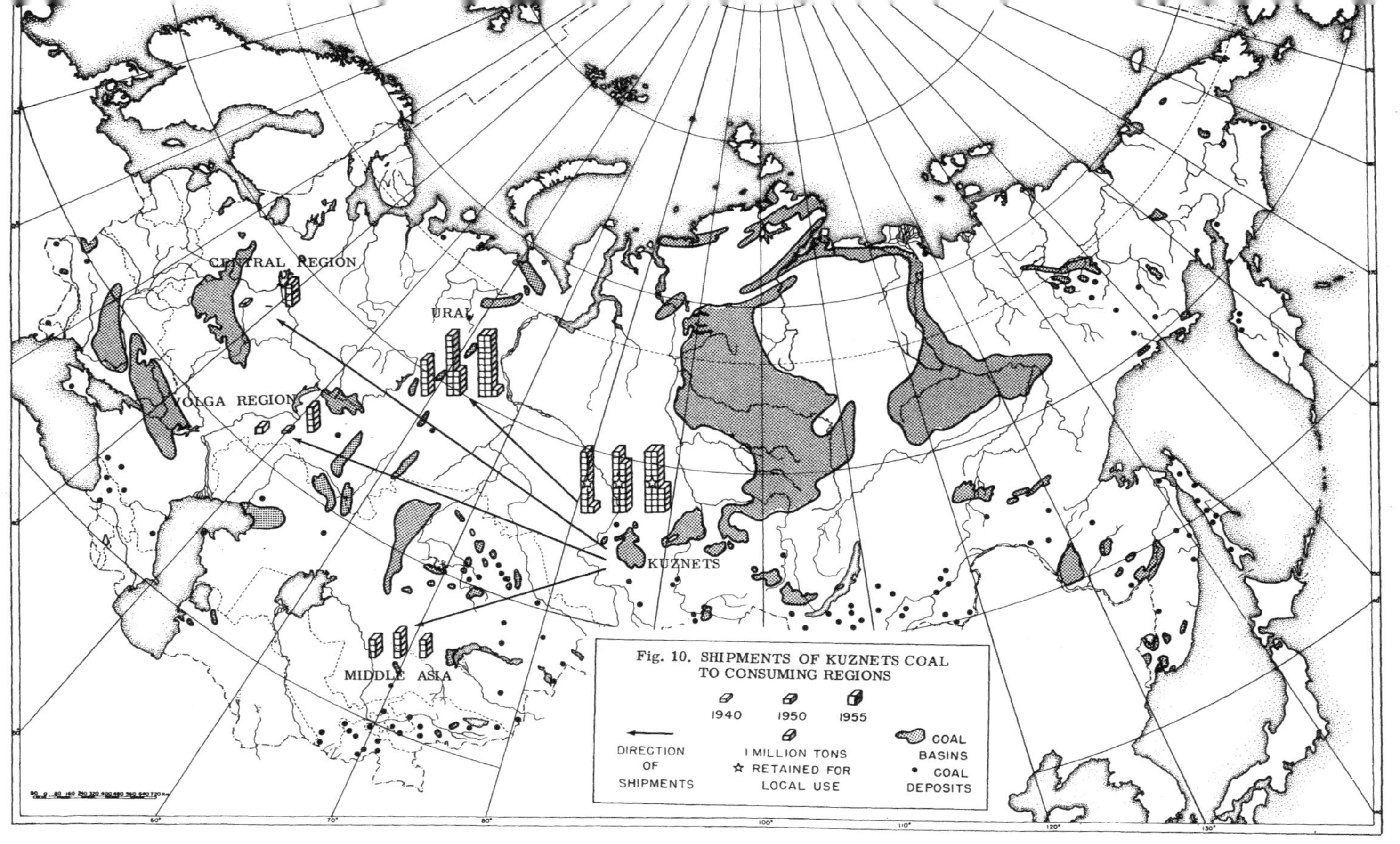

Fig. 10. SHIPMENTS OF KUZNETS COAL TO CONSUMING REGIONS

crease was primarily occasioned by the opening of the Karaganda and Kuznetsk Basins during the early Five Year Plans.

The Soviets have done much to develop their low quality, low energy content brown coal as local fuel in the European regions. These fuels are unable to meet the overall energy requirements of this region, and are unsuitable for many industrial processes. However, they are effective for generating power in thermoelectric stations. As industry expands in the European section, larger amounts of high quality, high energy fuel will be required from the eastern basins. Thus, long-distance shipments of coal may become a stable feature of the Soviet economy. A contradictory situation exists between Soviet theory, which assigns the consumer to a location adjacent to the source of raw materials and fuel, and reality, which at the current stage of development, finds the two antipodean in location. To abolish this contradiction, future developments in industrial expansion should occur east of the Urals. Higher quality coals and lower mining costs in the eastern basins are an inducement toward this development.

Low Cost Production in High Energy Potential Regions.—Cost data available on Soviet mining indicate that eastern basins can be mined more advantageously than the European ones. Comparative ruble costs for the mining of a ton of coal are presented in Table 7. If these costs are considered as index figures, then it is possible to arrive at valid conclusions concerning cost as a factor in the future geographical location of Soviet coal mining activity.

The most economically feasible region for the mining of all types of coal is the South Siberian Belt. Both high quality, high energy content hard coals and brown coal are inexpensive to mine here. A ton of Kuznetsk coal can be mined for a third less than the cost of mining a ton of Donets or Kizel hard coal, and its quality is better. M. I. Pomus contributed to an understanding of this cost advantage when he wrote of the Kuzbas that the: . . . "Average thickness of the layers is six times greater than in the Donbas." [35] Cheremkovo hard coal, can nevertheless be mined at less than a fourth the cost of Donets or Kizel coal, and for half the price of Karaganda coal.

Brown coal of the Kansk-Achinsk Basin is of better quality than its counterpart in either the Moscow Basin or the Dneper Basin. The Itatsk Deposit is the least expensive to mine. It is possible to mine 11.3 tons of Itatsk brown coal at a cost comparable to mining one ton of brown coal by the Moscow Coal Combinat. When Itatsk coal is compared on the basis of the thermal value obtained at the cost of mining one ton of coal in other basins, it still retains a significant advantage. For example, Donets hard coal has a thermal value of 7 million calories per ton. Here it costs approximately 95 rubles to mine these 7 million calories. This same 95 rubles expended in the Itatsk deposit would produce fuel with a thermal value of 47.5 million calories.

TABLE 7

THE COST OF PRODUCING ONE TON OF COAL IN VARIOUS BASINS AND DEPOSITS IN THE U.S.S.R. 1954, 1956, AND 1957 (Rubles and Kopeks)

Basin or Combinat	Type of Coal	1954	1956	1957
European Russia				
Donets Basin	Hard	93[a]	95.37[c]	
Moscow Basin				
Moscow combinat	Brown	---	56.73[c]	
Tula combinat	Brown	---	62.91[c]	
Urals				
Kizel Basin	Hard	---	91.09[c]	
Chelyabinsk combinat	Brown	---	47.40[c]	
Kazakhstan				
Karaganda combinat	Hard	47[a]	42.34[c]	
South Siberian Belt				
Kuznetsk Basin (shaft)	Hard	63[b]	56.60[c]	
Kuznetsk Basin (open pit)	---	---	29.00[d]	
Kansk-Achinsk Basin (open pit)				
Nazarov deposit	Brown	17[a]	---	11.82[g]
Irsha-Borodinsk	Brown	8[a]	6.70[d]	9.12[g]
Itatsk deposit	Brown	---	4.98[e]	
Cheremkovo Basin				
(open pit)	Hard	---	20.00[d]	20.7[h]
(shaft)		---	---	49.4[h]
Arctic – Subarctic				
Siberia				
Lena Basin				
Sangar Deposit	Brown	107.30[f]	---	107.3[g]
Kangalas deposit	Brown	74.36[f]	---	74.4[g]
Dzhebariki deposit	Hard	118.30[f]	---	
Tiksi Region				
Soginsk deposit	Brown	54.90[f]	---	
South Yakutsk Basin	Hard	75.[f]	---	
The North East				
Zyryansk Basin	Hard	84.90[f]	---	84.90[g]
Transbaikal				
Bukachachan deposit	Hard	72.[f]	---	69.44[g]

[a]Prof. V. Vasyutin, "Razmeshchenie Proizvoditelnykh Sil i Razvitie Khozyaistvo Vostochnykh Raionov Strany," *Pravda*, No. 14 (13807), 23 May 1956, p. 2.

[b]M. P. Ravich, *Toplivo v Shestoy Pyatiletka* (Moskva: Gospolitizdat, 1956), p. 19.

[c]D. Onika, "Economic Councils and the Problems of Manifold Uses of the Resources of Economic Regions," *Problems of Economics*, Vol. 1, No. 1, English translation of *Voprosy Ekonomiki*, p. 20.

[d]L. Semenov, "Ugolnye Resursy Krasnoyarskogo Kraya i Perspektivy Ikh Ispolzovaniya," *Planovoe Khozyaistvo*, No. 12 (1957), 78.

[e]L. V. Skripka, *Itatskoe Ugolnoe Mestorozhdenie* (Moskva: Ugletekhizdat, 1957), p. 57.

[f]V. F. Vasyutin, ed., *Problemy Razvitiya Promyshlennost i Transporta Yakutskoi A.S.S.R.* (Moskva: Akademii Nauk S.S.S.R., 1958), pp. 124-143.

[g]Akademiya Nauk S.S.S.R., *Razvitie Prozvoditelnykh Sil Vostochnoy Sibiri, Toplivo i Toplivnaya Promyshlennost* (Moskva: Izdatelstvo Akademii Nauk S.S.S.R., 1960), p. 7.

[h]*Ibid.*, p. 67.

Itatsk coal has neither the versatility nor the thermal value per unit of weight of Donets coal, but its use as a local fuel for the generation of electric power would be extremely inexpensive.

Karaganda can produce two tons of high energy content coal at the same price for which one ton is produced in either the Donets or the Kizel Basins. Specific data are not available on the cost of mining brown coal at the Maikyubin deposit in Kazakhstan, but Onika, comparing it with Urals brown coal, states that: [36]

> ... the cost of one ton of Chelyabinsk coal, including transportation to Ufa, amounts of 125 rubles: of Maikyubin coal—108.31 rubles, or about 15% lower.

Impressively low production cost in these eastern basins explains the Soviet's ability to maintain long freight hauls to European consumers. These lower costs are undoubtedly a factor in maintaining the stability of long hauls.

Arctic and Subarctic Siberian mining is expensive. Firstly, there are the "Transpolar" wage rates, [37] which are higher than those for other regions of the country; secondly, there is the severe climate, which even requires blasting in open pit operations; [38] and thirdly, there is the lack of mechanization, plus the antiquity of the equipment used in the region. [39] Hard and brown coals of the Lena Basin are the most expensive to mine in the country. Tiksi on the Arctic Coast, however, mines brown coal at a lower cost than the Moscow Basin. Hard coal can also be mined with less expense in the South Yakutsk Basin than it can in the Donets or Kizel Basins.

The Zyryansk Basin in the Northeast region is able to produce high quality hard coal for less than any European Basin.

Transbaikal hard coal of coking quality is mined in the Bukachachan deposit at only three-fourth the cost of mining a ton of Donets coal.

The low cost of production in the high energy potential regions east of the Urals magnifies the present inefficient situation where most of the actual Soviet coal production takes place in the European region. Throughout Mr. Khrushchev's report to the 21st Congress, is the admission that the eastern areas have a greater potential which could be developed at lower cost. [40] Coal and other fuels are mentioned in particular in this speech.

Attempts to Aid Inefficient Regional Production.—Confronted with a situation in which coal production, consumption, and reserves are located poles apart, the Soviet Authorities have tried to correct this inefficient condition. Their endeavors within the framework of the coal industry include the underground gasification of coal, intensive geological exploration for new coal deposits in the European region, and concentrated study of the area east of the Kuznetsk Basin.

Underground gasification of coal had its inception in 1935 with the formation of the "Trust Podzemgaz." Construction of an industrial station at Lisichansk in the Donets, and an experimental industrial station, Podzemgaz, near Tula in the Moscow Basin, began in 1938. [41] Lisichansk was placed in operation in December, 1940, and the station Podzemgaz, near Tula, was commissioned in November, 1940. Production was discontinued in both stations the following year by the invasion of the German Army.

Short, but satisfactory performances were given by the two stations. Lisichansk produced 37.4 million cubic meters of gas, with a calorie content of 1,100 calories per cubic meter during its eight months of operation, and the Podzemgaz station produced 47.5 million cubic meters with a calorie content of 830 calories per cubic meter during its twelve months of operation.

Early post war developments in underground gasification of coal consisted of the reconstruction and refurbishing of the Donets and Moscow stations. Both were in full production by 1948. [42] Two additional experimental-industrial stations were constructed by 1955, the South Abinsk (Podzemgaz) in the Kuznetsk Basin, and the Shatskoy in the Moscow Basin. [43] (see Table 8). Construction began on the Angren Station, near Tashkent in Uzbekistan in 1952, [44] and gas was first produced here on an industrial basis in January of 1958. [45] Kamen, in the Donbas, was under construction at this time, and experimental work had begun on two projected gasification stations in the Moscow Basin and three in the Kuznetsk.

Thermoelectric stations consume almost the entire output of gas from underground generators. Fully automatic gasification stations operating simultaneously with thermoelectric generating stations are eventually planned. At present, electricity produced from gas procured by underground generation is exceedingly insignificant. Production of electricity in this manner amounted to just 0.05 per cent of the total in 1955 (see General Appendix Table I). The new Seven Year Plan reveals that gas generated underground will comprise but 0.5 per cent of the fuel balance for all regions east of the Urals in 1965. [46] Eastern coal deposits considered expedient for the underground gasification of coal are the Itatsk and Nazarov in the Kansk-Achinsk Basin, and in the Ubagan (Turgay) Basin. [47]

Table 8 suggests that great differences exist between stations. Stations with larger capacities, projected for the Kuznetsk Basin, are credited with the ability to produce gas at less expense. All of the established stations have a higher cost factor and a lower heating value for the gas generated than do the projected stations. The heating values of gas generated in established stations can be explained by the fact that they are gasifying brown coal, while the projected stations in the Kuznets Basin will be operating on hard coals. Es-

TABLE 8

AN ECONOMIC AND TECHNICAL ANALYSIS OF UNDERGROUND GASIFICATION

Categories of analysis	Donets Basin		Moscow	
	Lisichansk E	Kamen C	Podzemgaz E	Shatskoy E
Capital investment in rubles 10^6	67.6[a]	41.0	82.7[b]	47.3
Gross production of gas in cubic meters 10^6	112.3[a]	365.0	401.8[a]	662.4
Production of commodity gas in cubic meters 10^6	89.7[a]	---	359.4[a]	633.6
Calories per cubic meter of gas	838.0[a]	900.0	770.0[a]	900.0
Capital investment in the production of one mega-calorie of fuel, rubles per year	---	125.0	251.0[b]	79.0
Number of workers (total)	480.0[a]	227.0	505.0[a]	450.0
Number of workers (laborers)	349.0[a]	141.0	374.0[a]	353.0
Production of gas per worker (total) per month in cubic meters 10^3	19.6[a]	134.0	66.3[a]	123.0
The cost of 1,000 cubic meters of gas in rubles	80.7[b]	21.80	37.19[b]	21.95
The cost of one mega-calorie of fuel in rubles	---	24.3	46.5[b]	25.6

1. Compiled from G. D. Bakulev, *Ekonomicheskiy Analiz Podzemnoy Gazifikatsii Ugley*, (Economic Analysis of the Underground Gasification of Coal) (Moskva: Akademii Nauk S.S.S.R., 1957), pp. 29, 33, 35, 57, 60, 62, 63, 64, 81, 86.

E — established fully operating plants.

C - under construction — completion date 1960.

tablished stations, however, had lower initial capital investments. Larger numbers of workers are utilized in existing stations than are planned for the projected stations. This may be accounted for, in part, by the fact that they are also experimental stations.

Coal gasified underground is an experimental attempt to rationalize the fuel base in the U.S.S.R. by utilizing coal which might otherwise be considered unmineable, and thus curtailing long freight hauls and diverting higher grades of coal in a given region to more efficient uses. The practicality of this technique will remain unknown to

STATIONS, ESTABLISHED, UNDER CONSTRUCTION, AND PROJECTED(1)

Basin		Kuznetsk Basin				Middle Asia
Gryzlov	North Tula	South Kuzbas	Stalinsk	South Abinsk[c]	South Abinsk	Angren
P	P	P	P	E	P	E
173.3	71.5	312.8	258.7	21.0	153.0	128.2
3,140.0	493.0	7,990.0	3,995.0	582.5	4,575.0	2,320.0
2,770.0	484.0	7,012.0	3,570.0	539.5	4,250.0	1,740.0
900.0	900.0	1,200.0	1,200.0	1,100.0	1,200.0	1,000.0
61.3	161.1	32.6	54.0	32.8	28.0	55.3
500.0	300.0	920.0	660.0	234.0	640.0	626.0
371.0	190.0	740.0	498.0	---	471.0	508.0
524.0	146.4	724.0	504.0	206.7	595.0	322.0
10.76	25.80	6.1	6.70	20.09	6.10	18.42
11.9	28.7	5.1	5.58	15.35	14.0	18.42

P – projected construction – completion after 1960 date unknown.

[a]data 1956.

[b]data 1955.

[c]This station is also known under the name of Podzemgaz.

scholars outside the Soviet Union until the Soviets reveal exactly how much gas is produced per ton of coal. Evidence to this effect has not been uncovered by the author. Bakulev, however, does present the total gas production of two stations in relation to the total coal which they burned underground. When ratios were computed from this data, the calorie content derived from a ton of coal burned underground equalled, or nearly equalled, the calorie content for that specific type of coal. This is an improbable occurrence because coke remains underground in this type of operation and some of the heat value is lost. For this reason, the writer must conclude that the calorie contents per cubic meter of gas presented in Table 8 are maximum values obtained under test conditions.

Gasification of coal underground may alleviate the coal shortage in European Russia, but Table 8 demonstrates that it is the high energy content eastern regions that will be the most efficient producers of this product. Regardless of the economy effected by this technique, the European region must search elsewhere for fuel.

Geological exploration in post war years revealed three additional sources of coal in European Russia: the Lvov-Volyn Basin in the Western Ukraine, the Greater Donbas, and the accidentally discovered Kama Basin. The first two are now being exploited.

The Lvov-Volyn Basin has been rapidly developed with the expectation that it will relieve the European West of its recurring coal shortage; national attention was focused throughout 1956 on this development. [48] Zastavnyy shows the expectations of the basin with two maps in his book on its economic-geographical characteristics. His plan calls for the shipment of coal from the Lvov-Volyn Basin to Leningrad, the Baltic States, Moldavia, and the Central Industrial Region. This would allow the Donets to ship greater amounts of coal to the Transcaucasus and the Volga Regions. [49]

The Lvov-Volyn Basin contains five workable deposits with coal seams averaging 0.7 to 1.2 meters thick. [50] Coal in this basin is similar to the gassy (G) coal of the Krasnoarmeisk section of the Donbas. While the ash content is relatively high and fluctuates widely, it nevertheless has a lower ash content than coals which are set aside in other basins for underground gasification. [51] In 1956, five operating mines with a capacity of 1,650 thousand tons produced 522.6 thousand tons of coal. By 1958 Lvov mines produced 1.94 million tons. Scheduled for construction by 1960, are an additional 16 mines with a total capacity of 8.4 million tons annually. [52]

It is difficult to conceive of this basin as a major exporter of coal, when its ultimate capacity is destined to be less than the annual production of the Dneper brown coal basin. Even though the quality and energy content of its coal may be higher, the actual contribution in tonnage to be made each year is not great, and will diminish in significance as consuming industries expand. Without it, however, a greater burden would be placed upon railroad facilities. Geographically, this basin may be the major factor in the location of new industrial enterprises of a coal-chemical nature. [53]

Intensive geological exploration, conducted since 1946, has expanded the boundaries of the Donets mining areas to the West, North, and East. [54] A reiteration of European Russia's reliance on the Donets Basin is unnecessary. It is sufficient to indicate that as the boundaries are enlarged from year to year, new mine construction is rapidly instituted. [55] In 1955, the Donets Basin contained 55.2 per cent of the nation's mines. Listed below are figures showing the actual increase in the number of mines in this basin.

Year	Number of Mines[a]
1940	311
1950	464
1955	512
1956	580

[a]Peredukhin, op. cit., p. 8, 1940-1955. Troyanski, *op. cit.*, p. 193, for 1956.

Recently constructed mines have a greater daily productive capacity than those built in previous years. Increased Donets production contributes to a greater efficiency in the coal industry in European Russia. Coal mined there is expensive, but high in energy content and versatile in use. Its utilization negates long-distance shipments from other basins.

Soviet planners are directing the construction of new mining and industrial enterprises eastward in a positive approach toward the rationalization of industry. In August of 1958, a conference on the "Development of the Productive Forces of Eastern Siberia" was convened at Irkutsk. Over 8,000 persons attended, and more than 800 reports were presented. [56] Continually stressed throughout the conference was the existence of large reserves of high quality coal that could be mined cheaply in this region. The noted authority on Soviet energy resources, A. E. Probst, summarized the substance of many of these reports with the following statement quoted in Pravda:

> The total geological reserves of coal in this region according to the last year's calculations, amount to nearly 7 trillion tons; this constitutes about 80 per cent of the reserves of the U.S.S.R. Resources of coal in Eastern Siberia are approximately three times greater than in the U.S.A., and considerably more than in all the capitalistic countries put together. No less important a circumstance is the fact that cost of producing coal here is approximately 10 times lower, than the average cost in the Union for underground mining. [57]

Other reports outlined the feasibility of constructing a new ferrous metallurgical center at Taishet, [58] and called for an integration of the iron ores of the Angaro-Pitsk Basin with Minusinsk coal for the creation of the "Central Krasnoyarsk Industrial Complex." [59]

Many of the essential ideas of these reports are found incorporated in the Seven Year Plan for 1959-1965, and others are expressed in programs projected for 1975. Future industrial centers capable of effecting a rational union between the high energy content coal of the east and its multitudinous resources are in the initial stages of creation. Effective culmination of these plans could result in a more efficient production and consumption of coal than that which now exists in European Russia. This will not, however, alleviate or revise the traditional problem confronting this region.

SUMMARY

Coal production has progressively increased during the Soviet regime. Characteristic of this growth in production is a decreasing dependence on coal of the Donets Basin, and an increasing use of brown coal in the Soviet economy. Both trends reflect Soviet attempts to utilize all fuel resources. Even though the Soviets have achieved some success in shifting production eastward to the high energy reserve basins, production and consumption remain centered in the European regions. Present production and consumption patterns are geographically opposite in location to high quality, high energy content reserve areas. Static consumption patterns have stabilized and aggravated long-distance coal shipments.

These features of the fuel economy are contrary to the Communist doctrine on the rationalization of industry. Moreover, lower production costs in basins with high quality, high energy content coals magnify the present nonrational situation. Ameliorative attempts have resulted in the discovery of additional sources of coal in current high cost producing regions in European Russia. The contemplated gain from these discoveries may only be punitive when related to existing rates of industrial expansion in the area. In the opinion of this student, the one positive approach adopted by the Soviets is the present orientation of planning toward development of the eastern basins.

NOTES – CHAPTER II

1. G. D. Bakulev, *Voprosy Ekonomiki Topliva v S.S.S.R.* (Moskva: Gospolitizdat, 1957), p. 6.

2. V. I. Lenin, *Collected Works*, Vol. 30, p. 461.

3. E. S. Sofiev, ed., *Dostizheniya Sovietsky Vlasti As Sork Let V Tsifrakh, Statisticheskiy Sbornik* (Moskva: Gospolitizdat, 1957), p. 227.

4. See General Appendix Table I.

5. N. G. Feitelman, *Sebestoimost Uglia i Puti ee Snizheniia* (Moskva: Ugletekhizdat, 1956), p. 15.

6. Prof. N. Melnikov, "Bystree Razvivat Dobychu Ugliya Otkrytum Sposobom," *Pravda*, No. 321 (13254), 17 November, 1954, p. 2.

7. Lynn Turgeon, "Cost-Price Relationships in Basic Industries During the Soviet Planning Era," *Soviet Studies*, IX, No. 2 (Oct. 1957), 152-153.

8. N. G. Feitelman, *op. cit.*, p. 16.

9. I. A. Kulev, *op. cit.*, p. 8.

10. Anon., "Rezko Ovelichit' Dobychu Uglya Donbasse," *Pravda*, No. 293 (13956), 19 Oct., 1956, p. 2.

11. Peredukhin, *op. cit.*, p. 6.

12. V. P. Aksenov, N. O. Zamorenov, and A. D. Rybin, *Razrabotka Burykh Ugley Ukrainy (Dneprovskii Bassein)* (Kiev: Gos. izd-vo tekh. lit-ry, 1955), pp. 5-6.

13. *Konferentsiia Po Kompleksnomu Razvitiin Buro-Ugolnoi Promyshlennosti Ukrainskoi S.S.R.* (Kiev: 1948), 388 p.

14. P. S. Bulgakov, ed., "Tsentralne Statistichne Upravlinnya pri Radi Ministriv S.S.S.R Statistichne Upravlivnnya Ukrainskoi R.S.R.," *Narodne Gosprodarstvo Ukrainskoi R.S.R. Statistichnii Zdirnik* (Kiev: Derzhstatvidav, 1957), pp. 37-38.

15. I. G. Roshchupkin, *Razrabotka Ugolnykh Plastov v Podmoskovnom Basseine* (Moskva: Ugletekhizdat, 1956), p. 3.

16. D. G. Onika, *Podmoskovnyi Ugolnyi Bassein 1855-1955* (Moskva: "Moskovskiy Robochiy," 1956) p. 30.

17. I. G. Roshchupkin, *op. cit.*, p. 13.

18. V. Vityazeva, "O Razvitii Pechorskogo Ugolnogo Basseina," *Planovoe Khozyaistvo*, No. 11, (Nov. 1957), 80-81.

19. A. A. Chernov, "Nesmetnye Bogatstba Severa," *Pravda*, No. 205 (14234), 23 July 1957, p. 2.

20. P. N. Stepanov, *op. cit.*, p. 12.

21. Y. G. Makushkina, A. M. Marinvoa, and E. I. Rassadnikova, *Energetika Urala za 40 Let* (Moskva: Gosenergoizdata, 1958), p. 22.

22. Ya. G. Makushkina, *op. cit.*, p. 20.

23. Abil-Mazhit Tastenov, *Sloevaya Razrabotka Moshchnykh Ugolnykh Plastov v Karagandinskom Basseine* (Moskva: Ugletekhizdat, 1956) p. 7.

24. A. Zademidko, *op. cit.*, p. 2.

25. V. Kozhevin (Chief of Kuznetsk Coal Combinat), "Kuznetskiy Bassein v Novoy Pyatiletke," *Pravda*, No. 76 (13739), 16 Mar., 1956, p. 2.

26. L. D. Shevyakova, ed., *Razrabotka Krutopadayuskchikh Plastov Kuzbassa* (Moskva: Ugletekhizdat, 1956), p. 93.

27. S. Kuryshkin, "Razvivat Otkytuyu Dobychu Uglya v Kuzbasse," *Pravda*, No. 354, (13652), 20 Dec., 1955, p. 2.

28. V. Kozhevin, *op. cit.*, p. 2.

29. S. Seryy, "Za Dalneishee Razuitie Ugolnoy Promyshlennosti Sakhalina," *Pravda*, No. 52 (13350), 21 Feb., 1955, p. 2.

30. V. Vityazeva, *op. cit.*, p. 81.

31. S. S. Balzak, V. F. Vasyutin, and Y. G. Feigin, *Economic Geography of the U.S.S.R.*, (New York: The Macmillan Co., 1949), p. 214. (American edition edited by Chauncy Harris). In 1940 when the Russian edition of this book was published, the authors claimed that hundreds of thousands of tons of Kuznetsk coal were being shipped to the Far East and that this practice should cease. Data in Kombinat Kuzbassugol Kemerovskovo Sovnarkhoza, *Kuznetskiy Ugolnyy Bassein Statiticheskiy Spravochnik* (Moskva: Ugletekhizdat, 1959), p. 366 indicates the shipment of one thousand tons in 1956.

32. R. S. Livshits, *Ocherki Po Razmesheniyu Promyshlennosti S.S.S.R.* (Moskva: Akademii Nauk S.S.S.R., 1954), pp. 6-58; and Harry Schwartz, *Russia's Soviet Economy* (Englewood Cliffs, N. J.: Prentice Hall, Inc., 1950), pp. 552-558.

33. N. N. Baranski, *Economic Geography of the U.S.S.R.* (Moscow: Foreign Languages Publishing House, 1955), p. 24.

34. Benjamin I. Weitz, ed., *Electric Power Development in the U.S.S.R.* (Moscow: 1953), p. 151.

35. M. I. Pomus, *Zapadnaya Sibir Ekonomiko-Geograficheskaya Kharakteristika* (Moskva: Geografgiz, 1956), p. 38.

36. D. Onika, "Economic Councils and the Problems of Manifold Use of the Resources of Economic Regions," *Problems of Economics*, Vol. 1, No. 1, (English translation of Voprosy Ekonomiki), p. 20.

37. V. Vityazeva, *op. cit.*, p. 80.

38. L. Semenov, *op. cit.*, p. 76.

39. V. F. Vasyutin, *Problemy Razvitiya Promyshlennost*, *op. cit.*, p. 144.

40. N. S. Khrushchev, *Target Figures for the U.S.S.R.'s Economic Development from 1959 to 1965, Full Text* (Washington: Embassy of the USSR), Press Department No. 567, 1958, p. 26.

41. F. I. Kleimenov, *Podzemnai Gazifikatsiya Uglei* (Moskva: Izdatelstvo Znanie, 1955), Series IV, No. 35, p. 40.

42. G. D. Bakulev, *Ekonomicheskiy Analiz Podzemnoy Gasifikatsii Ugley* (Moskva: Akademii Nauk S.S.S.R., 1957), pp. 27-29.

43. N. Lavrov, "Usemerno Razvivat Gazovuyu Promyshlennost," *Pravda*, No. 145 (13806), 22 May, 1956, p. 12.

44. F. Kleimenov, "Bolshe Unimaniya Razvitiyu Podzemnoy Gazifikatsii Ugley," *Pravda*, No. 265 (13923), 21 Sept., 1956, p. 2.

45. *Pravda*, No. 31 (14425), 31 January, 1958, p. 1.

46. Yu. Bokserman, V. Kalamkarov, A. Kortunov, "Zadachi Razvitiya Gazovoy Promyshlennost," *Planovoe Khozyaistvo*, No. 12 (1958), 34.

47. M. M. Altshuler, R. E. Leshchiner, and E. Yu. Chernyak, "Razvitiya Podzemnoy Gazifikatsii Ugley v Turgaiskom Basseine," *Podzemnaya Gazifkatsiya Ugley*, No. 3 (1957), p. 45.

48. "Soveshehanie Gornyakov Lvovsko-Volynskogo Ugolnogo Basseina," *Pravda*, No. 234 (13897), 21 Aug., 1956, p. 1. – A. Chernichenko and A. Bogma, "Soveshchanie Shakhtostroiteley Lvovsko-Volynskogo Ugolnogo Basseina," *Pravda*, No. 235 (13898), 22 Aug., 1956, p. 1. – "Semya Shakhterov," *Pravda*, No. 316 (13979), 11 Nov., 1956, p. 1.

49. F. D. Zastavnyy, *Lvovsko-Volynskyi Vukilnyi Bassein, Ekonom-Geograficheskyi Narya* (Lvov: Knyshkovo-Zhurnal'ne Vyd-vo, 1956), pp. 57-59.

50. M. I. Struev, "Uglenost Lvov-Volynskogo Kamennovgolnoko Basseina," *Izvestiya Dnepropetrovsk Gorn Institute*, No. 11, (1957), 118-132.

51. V. I. Dal, O. S. Fomenko, and G. C. Anenkova, "O Prigodnosti Uglei Lvovsko-Volynskogo Mestrozhdeniya Dlya Tseley Gazifikatsuua," *Gazovaya Promyshlennost*, No. 1 (1957), 11.

52. *Ugol Ukrainy*, No. 11 (1957), pp. 1-22.

53. Zastavnyy presents a strong case for the development of a chemical industry in this region in his article, "The Lvov-Volyn Basin and the Prospective Development of a Chemical Industry in the Western Oblasts of the U.S.S.R.: (Ukraine)" *Dopovidi t Apovidomleniya, Lvivsk Universitit*, 1957, Bin. 7, Chasti 3, pp. 23-27.

54. V. S. Popov, "Bolshoy Donbass," *Ugol Ukrainy*, No. 8 (1957), 23-27.

55. "Donbass Mozhet i Dolzhet Davat Todine Bolshe Ugiya," *Pravda*, No. 231 (13894), 19 Aug., 1956, p. 1.

56. V. Nemchinov, "Perspektivy Razvitiyu Proizvodietalnykh Sil Vostochnoy Sibir," *Planovoe Khozyaistvo*, No. 11 (Dec., 1958), 9.

57. A. Azizyan, N. Pecherskii, M. Chernenko, "Vostochnaya Sibir Stanat Novym Moshchym Centrom Industrii," *Pravda*, No. 232 (14626), 20 Aug., 1958, p. 1.

58. V. Nemchinov, *op. cit.*, p. 11.

59. A. Zubov, "Problemy Razvitiya Tyazheloy Promyshlennosti v Krasnoyarskom Krae," *Planovoe Khozyaistvo*, No. 12 (Dec., 1958), 65-66.

Chapter III

OIL SHALE: THE DISTRIBUTION OF RESERVES, ENERGY POTENTIAL, AND PRODUCTION

Geographic location is the significant aspect of oil shale in the energy balance of the Soviet Union, not quantity or quality. In the localities where it occurs and is developed, it is vitally important as a local fuel. Estonia and neighboring Baltic areas, which lack other types of fuel resources, are especially dependent on oil shale in their energy economy. To a lesser degree this is true of the Middle Volga Region.

Geological reserves of oil shale in the U.S.S.R. reportedly total 156 billion tons, an insignificant amount when compared to other energy resources in the Soviet Union (Shabarova and Tyzhnova, pp. 162-176). These reserves are less than half of the estimated geological reserves of coal in the Pechorá Basin. However, surveys for this type of fuel are not complete, and much remains to be accomplished in remote regions of Siberia.

Mineable reserves of 54.5×10^9 tons are roughly equal to the mineable reserves of brown coal in the Kazakh Republic (see Shale Appendix - Table I). Shale classified as mineable by the Soviets has an energy potential of 120.5×10^{12} kilowatt hours, which approximates the energy potential of the 15.9×10^9 tons of hard coal in the Gorlov deposit of Western Siberia. These mineable reserves of oil shale are sufficient to last 4,999 years at the 1955-1956 rate of production.

The mining of oil shale increased 733 per cent from 1945 to 1956. A small part of the growth can be credited to the reconstruction of war damaged areas; however, an intense development effort in the Baltic Basin has been responsible for the major increase. In all probability, mining has been concentrated on the higher class of shale.

THE SIGNIFICANCE OF SHALE IN THE ENERGY BALANCE OF THE SOVIET UNION

Oil shale is the least important fuel in the balance of consumed energy resources in the Soviet Union. In 1958, it accounted for 0.8 per cent of the energy consumed. [1] According to the projected plans, oil shale will supply but 1.1 per cent of the fuel consumed in the energy balance in 1965. At that time, wood fuel will account for 3.5 per cent of the energy consumed.

Bakulev stated in his book, *Voprosy Ekonomiki Topliva V S.S.S.R.*, that shale comprised 0.9 per cent of the total energy potential of the country when all energy resources were measured in conventional fuel units. [2] Since this book was published in 1957, new data on the total reserves of oil shale and coal have been announced. This data reveal that oil shale reserves have increased by 192 per cent, coal reserves have increased by 426 per cent, and oil reserves by 820 per cent. The actual position of shale in the energy balance of the Soviet Union is much less than that claimed by Bakulev.

Shale is not significant on a national scale, but it is important on a local or regional level. In the Baltic States, and in Estonia in particular, oil shale is a valuable part of the energy economy. In an interview on May 18, 1959, Dr. Valdar Jaanusson, formerly of the Geological Department, Tartu University, in Estonia, and presently a member of the Paleontological Institute of Uppsala University, informed the author that Estonian shale is so high in oil content that it is utilized as a solid fuel in many phases of the economy. He stated that this shale, which "may be ignited by holding a match to it," is burned in its solid form in railroad locomotives and power generators in plants. A heavy tar residue (bitumen), which makes boiler cleaning a frequent necessity, is the main disadvantage of its use in this form. Dr. Jaanusson also stated that the ash content of Estonian shale was less than Moscow Basin brown coal. Under the Soviet regime, mining of this shale by shaft and open pit methods had increased production tremendously and thus increased the position of shale in the Baltic energy economy.

CLASSIFICATION OF SHALE

The classification of oil shale is done on the basis of its calorie content and per cent of tar. [3] Energy or power shale must have an absolute minimum calorie content of 1,700 calories per kilogram; little shale is reported with a calorie content this low. Technological shales, those apparently reserved for processing for the Soviet's higher grade fuels, chemical by-products, and medical derivatives, must have a minimum tar content of at least 10 to 12 per cent.

Unfortunately, the Soviet government does not report reserves according to these classifications. Reserve data are reported by tonnage; thus the quality of the reserves, more often than not, must be collected from multiple sources. Therefore, it is difficult to establish precisely how much shale should be credited to either category. Data in Shabarova and Tyzhnova indicate that shale of the highest quality is found in the Baltic Basin. Kokhtla-Yarve, in the Estonian section of this basin, is the site of the only integrated processing plant; so by inference it may be assumed that the major portion of the technological reserves is located here. Some processing does occur in the Saratov area, but it is subordinate to the major objective, that of producing crude oil as fuel for thermoelectric stations. Shale classified as energy fuel is reduced to a crude oil product, which is burned not only in thermoelectric stations, but by railroad engines as well.

GEOLOGICAL AND MINEABLE RESERVES OF SHALE

Oil shale reserves in the Soviet Union are located in five widely dispersed regions (see Fig. 11). European Russia contains two of these regions, the Baltic, and the East and North Russian Platform. Kazakhstan, with one large deposit and several small deposits, is a third region. Shale of the Kuznetsk Basin composes the fourth region. Dominating all of these deposits in tonnage, is the fifth and largest region, the Northeast Siberian Platform. The geological and mineable reserves of oil shale are distributed among these basins as follows:[a]

Region	Per Cent of Geological Reserves	Per Cent of Mineable Reserves
Baltic Basin	12.8	25.9
North and East Russian Platform	11.5	31.2
Kazakhstan	2.6	1.3
Kuznetsk Basin	1.3	0.9
Northeast Siberian Platform	71.8	40.7
	100.0	100.0

[a]Computed from data in Shabarova and Tyzhnova and Shale Appendix Table I. Unless otherwise noted, all shale listings are adapted from these sources.

Of the 156 billion tons of oil shale listed as geological reserves in these five regions, only 4.0 per cent are proven reserves. Probable reserves account for 15.3 per cent of the total, and possible reserves 80.7 per cent. Over 71 per cent of the possible reserves are located

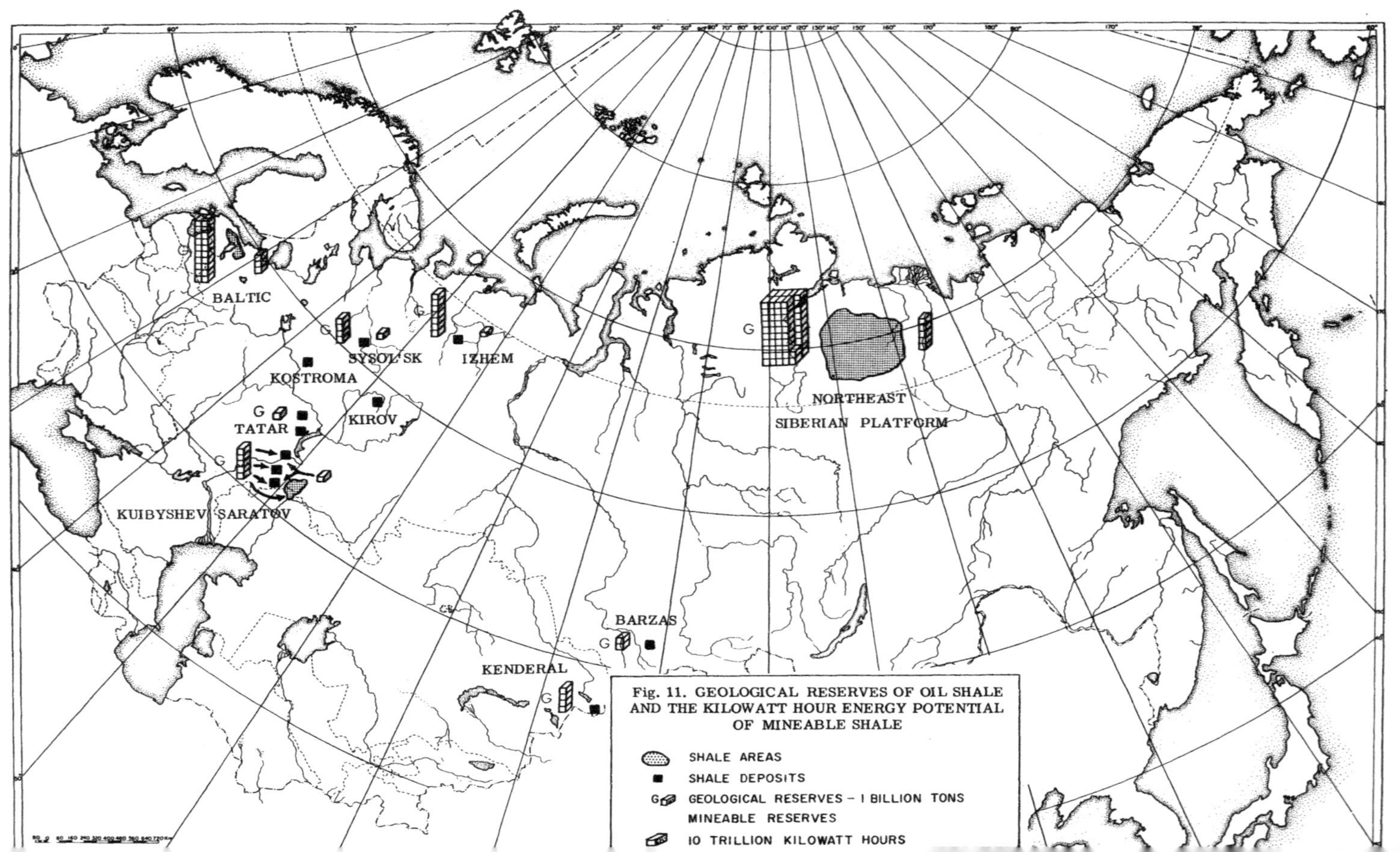

Fig. 11. GEOLOGICAL RESERVES OF OIL SHALE AND THE KILOWATT HOUR ENERGY POTENTIAL OF MINEABLE SHALE

in Northeastern Siberia, which indicates the need for more detailed surveys in that region. The Baltic Basin has a larger per cent of proven reserves than any other basin.

The Baltic Basin.—The Baltic Basin is divided into three sections: the Estonian, the Gudov, and the Chudov. Since the division is political, not geological, the Gudov and Chudov are both a continuation of the Estonian deposits (see Fig. 9). A preponderance of the geological and mineable reserves are located within the Estonian S.S.R., as the following listing indicates. Quality of shale in this basin is superior to that of any other basin in the Soviet Union. It contains less moisture and ash but a higher volatile and tar content than shales of other basins.

Baltic Basin Deposits	Geological Reserves 10^6	Mineable Reserves 10^6
Estonia	14,477	10,463
Gudov	5,242	3,551
Chudov	711	206
Total	20,430	14,220

The carbon content of the organic matter of oil shale, that has not been subject to weathering, ranges between 76.7 and 77.6 per cent. [4] Higher quality shales are concentrated in the center of the basin.

The North and East Russian Platform.—The crescent-shaped formation of the North and East Russian Platform arches eastward from the Komi A.S.S.R. toward the Urals, then southwestward to the Middle Volga Region. Within the confines of the region are eight individual oil shale deposits:

Russian Platform Deposits	Geological Reserves Tons 10^6	Mineable Reserves Tons 10^6
Izhem (Urals)	6,800	6,350
Sysolsk (Komi)	5,401	5,401
Kostroma (Center)	48	42
Kirov (Volga)	354	---
Chuvash (Volga)	15	---
Murdov (Volga)	1	---
Tatar-Ulyanovsk (Volga)	1,036	1,029
Kuibyshev-Saratov (Volga)	4,551	3,972
	18,206	16,794

Most of the deposits listed above contain mineable shale, or shale with adequately thick seams and a low ash content. Throughout the entire platform, the moisture and ash contents of the deposits are higher than the Baltic Basin. Tar and volatile contents have a lower minimum than Baltic shales. It is probable that most of the Platform shale is classified as energy or power shale, and utilized in thermoelectric stations. The original "Goelro" plan envisioned thermal stations in the Middle Volga Region based on this shale. [5]

Kazakhstan.—Six different deposits of oil shale occur in the Kazakh S.S.R. (see Shabarova and Tyzhnova and Oil Shale Appendix Table I), but only two have deposits of mineable quality. Kenderal, the larger, has geological reserves of 4.0×10^9 tons and mineable reserves of just 698.0×10^6 tons. Its ash and moisture content is higher than Baltic shales and the volatile content is lower. Baikhozhin, the other, has geological reserves of 16.8×10^6 tons and mineable reserves of 6.8×10^6 tons. Oil shale in this deposit has the highest sulfur and lowest volatile content of any mineable deposits in the U.S.S.R. Kenderal has the lower ash content.

The Kuznetsk Basin.—One deposit of oil shale is reported for the Kuznetsk Basin. Geological reserves total 1.5×10^9 tons and mineable reserves 500×10^6 tons. Half of the geological reserves are concentrated above 600 meters and the other half found at depths of 600 to 1,800 meters. Mineable reserves are all above the 600 meter depth.

Northeast Siberian Platform.—Remote and isolated, the Northeast Siberian Platform contains over 71 per cent of the geological reserves of oil shale, and over 50 per cent of the mineable reserves. All reserves are in the possible category; none is proven or even probable. Of the 111×10^9 tons of shale reported for the area, 22×10^9 tons are considered mineable. If, in the future, the oil is distilled from this shale, the action will undoubtedly be influenced by the fact that the shale also contains a "rather high content of vanadium and molybdenum per unit of rock." [6]

THE KILOWATT HOUR ENERGY POTENTIAL OF SOVIET OIL SHALE

Method of Computation.—The procedure for computing the kilowatt hour energy potential of oil shale is the same as that used in determining the coal energy potential in Chapter I. The problems in establishing a thermal base from which to make these computations are essentially the same as those for coal.

Soviet sources provide a wide variety of thermal values, measured in calories per kilogram, for oil shale, but only for the combustible mass of the shale. These values are determined by discounting all nonburnables in the sample material. Power technicians and economic planners in the U.S.S.R. ignore the inflated values in their work, and use as a thermal base, 1,900,000 calories per metric ton in all of their calculations. [7] The author assumes that the Soviet scientists have chosen the most accurately determined thermal values for their work; therefore, he has adopted this thermal base as a standard for all computations in this section of the text.

The Total Energy Content of Soviet Oil Shale Deposits.—Mineable reserves of oil shale in the U.S.S.R. contain a potential 120.5×10^{12} kilowatt hours (see Oil Shale Appendix - Table I). This amounts to an ultimate per capita energy potential of only 602×10^{3} kilowatt hours per person. [8] When the energy potential of oil shale is compared to that for coal, it represents but 0.2 per cent of the energy potential for mineable reserves of coal. Undeniably the amount is insignificant. However, no matter how small the total oil shale potential is, when compared with other solid fuel resources it does attain local prominence by the absence, in quantity, of other energy resources. An understanding of the importance of oil shale in the Soviet economy depends on a knowledge of the distribution of the energy potential of this fuel.

DISTRIBUTION OF THE KILOWATT HOUR ENERGY POTENTIAL OF MINEABLE OIL SHALE RESERVES

The distribution of the kilowatt hour energy potential of mineable reserves of oil shale differs radically from the distributional pattern of geological reserves (see Fig. 11). This is to be expected because variances exist between the quantity of geological reserves and mineable reserves of oil shale. No essential difference exists, however, between the distribution of mineable reserves and the distribution of their energy potential. It is evident from the following listing that European Russia, including the Urals, possesses 56.8 per cent of the energy potential, but only 24.3 per cent of the geological reserves. This example alone expresses the difference in the distribution of geological and mineable reserves.

Approximately 41 per cent of the energy potential of oil shale is located in remote Arctic Siberia. Geological reserves in this region

Basin or Region	Per Cent of Geological Reserves	Energy Content Mineable Reserves Kwt. Hrs. 10^{12}	Per Cent Energy Content Mineable Reserves
Baltic Basin	12.8	31.4	26.1
East and North Russian Platform	11.5	37.1	30.7
Kazakhstan	2.6	1.6	1.3
Kuznetsk Basin	1.3	1.1	1.0
Northeast Siberian Platform	71.8	49.3	40.9
Total	100.0	120.5	100.0

total about 72 per cent, but reserves of mineable quality are few, and as previously stated, none is proven. Generalized data may adequately depict the overall distribution of the energy potential of oil shale, but it is only by a closer look at the regions themselves, that the true geographic significance can be obtained.

The Baltic Basin.—Approximately 26 per cent of the total energy potential of oil shale in the U.S.S.R. is concentrated in the Baltic Basin. Political boundaries subdivide this basin into three regions: Estonia, Gudov, and Chudov in the Leningrad Oblast of the R. S. F. S. R. Geologically, the area is a single unit with an energy potential of 31.4×10^{12} kilowatt hours.

The "Estonian Donbas" is the dominant region, as the following listing of the energy potentials of these three subdivisions indicates:

Subdivisions of the Baltic Basin	Energy Potential as a Per Cent of the U.S.S.R. Total	Energy Potential as a Per Cent of the Basin Total
Estonia	19.2	73.6
Gudov	6.5	24.8
Chudov	0.4	1.6
Total Baltic	26.1	100.0

Tarmisto states ... "The shale basin of the Estonian S.S.R. is now the largest mining-industrial region in the Baltic, ...". [9] We might add that the only industrial region of any size in the Baltic states is based upon the energy potential of this shale deposit. In fact, it is doubtful if any other industrial region in the world, however small, has as its prime source of energy an oil shale base.

Gudov, the second largest division of the Baltic Basin, contains approximately 25 per cent of the energy potential, and nearly 7 per cent of the total oil shale energy potential of the Soviet Union.

Chudov, the smallest division, has but 1.6 per cent of the energy potential. This represents 0.4 per cent of the entire energy potential for oil shale in the Union.

The East and North Russian Platform.—Larger than all other regions containing oil shale in the Soviet Union, is the East and North Russian Platform, with 30.7 per cent of the energy potential. Several widely dispersed deposits contribute to the platform's 37.1×10^{12} kilowatt hour energy potential.

Most of the energy potential in the East and North Russian Platform is located in the northern section. Two deposits, the Izhem and Sysolsk, account for 69.7 per cent of the Platform's total potential. This is equivalent to 22 per cent of the entire oil shale energy poten-

Deposits of the East and North Russian Platform	Energy Potential as a Per Cent of the U.S.S.R. Total	Energy Potential as a Per Cent of the Platform Total
Izhem (North)	12.0	37.7
Sysolsk (Komi ASSR)	10.0	32.0
Kostroma (Center)	0.1	1.0
Tatar-Ulyanovsk	1.2	6.0
Kuibyshev-Saratov	7.4	23.3
	30.7	100.0

tial in the U.S.S.R. With the projected development and utilization of this shale, the construction of thermoelectric stations at Izhem and Syktyvkar to serve these cities is planned, as well as a third station based on Izhem shale, to be a source of power for the electrification of the Pechora Railroad. [10]

Kostroma, with but one per cent of the region's kilowatt hour energy potential, has hydrogeological conditions which make mining difficult. Its development is problematical.

Oil shale deposits in the Volga Region comprise 29.3 per cent of the Platform's energy potential, or 8.6 per cent of the Union's energy potential. Practically all of this potential is located in Kuibyshev and Saratov Oblasts. Prior to the discovery of oil and gas in this so-called "energy poor" region, shale was developed as a local fuel for use in electrostations at Syzran and Saratov. [11] This development originated with the GOELRO plan.

Kazakhstan.—The kilowatt hour energy potential of oil shale in Kazakhstan is approximately one per cent of total for this type of fuel in the Soviet Union. Over 99 per cent of this potential is currently located in the Kenderal deposit. Further surveys on the recently discovered Ubagan deposits will undoubtedly reveal shale of a mineable quality and thus necessitate a reappraisal of the energy potential in this region.

The Kuznetsk Basin.—Barzasskii is the only known deposit of oil shale in the Kuznetsk Basin. It contains an even one per cent of the energy potential of oil shale in the country or 1.1×10^{12} kilowatt hours. Future development is improbable because it has the lowest tar content of any deposit in the Soviet Union.

The Northeast Siberian Platform.—Over 40 per cent of the 120.5×10^{12} kilowatt hour energy potential of oil shale in the U.S.S.R. is located in the Northeast Siberian Platform. Geological reserves of oil shale in this region constitute 71.8 per cent of the nation's total. If further survey work determines that a larger proportion of these reserves can be classified as mineable, then the energy potential will increase proportionately. Expectations in this direction are justified when we consider the increase in coal reserves attributed to this region as a result of resurvey between 1937 and 1955.

CORRELATIVE SUMMARY ON POTENTIAL AND QUALITY OF RESERVES

Most of the energy potential and higher quality oil shales of the Soviet Union are located in European Russia, excluding the Ural and Caucasus Mountains. Oil shale of high quality and potential in European Russia is located in three regions: the Baltic, the Northeast, and the Middle Volga. Available data indicate that the highest quality shales occur in two of these regions, the Baltic and Volga. Both are deficient in other solid fuels. Oil shale, unlike coal which has the vast preponderance of its energy potential scattered throughout several large deposits east of the Urals, has only one major deposit in the east. This Siberian deposit, although large, is of low quality.

The total kilowatt hour energy potential for oil shale in the Soviet Union is equivalent to but 0.2 per cent of the energy potential for coal. Moreover, because of its low energy content per unit of weight, oil shale is considered a local fuel. In spite of these facts, the energy potential to be derived from oil shale has a significant role in the energy economy of the Soviet Union because of Soviet efforts in developing local fuels. An understanding of this significance is predicated on a knowledge of oil shale production.

REGIONAL PATTERNS OF PRODUCTION AND CONSUMPTION

Oil shale appears unimportant and relatively insignificant when its role as an energy producer is considered in relation to the total energy economy of the Soviet Union. In 1955, electrical energy produced from oil shale amounted to 0.77 per cent of the Soviet total, or 6.6 kilowatt hours per person. Considerations of oil shale at this scale are worthwhile from an economic viewpoint, but they neglect the geographical significance of the product.

Shale is produced and utilized in areas considered deficient in other types of energy resources. This use of a local resource obviates the necessity for inordinately long freight hauls. Regions in which shale is currently and efficiently produced do exactly this for the Soviet economy.

The Position of Shale in the Energy Economy of Producing Regions.—Shale, as a local fuel, is vital to the energy economy of two regions, the Baltic and the Middle Volga. No other region has produced shale in the U.S.S.R. since 1956.

The Baltic Region.—Estonian deposits dominate the shale industry of the Baltic and contribute substantially to that Republic's economy. Nearly 78 per cent of the shale mined in the Baltic Basin originates in Estonia. This republic mines 6.4 tons of shale per person compared with 3.1 tons of coal per person in the Ukraine, and 1.9 tons of coal per person for the entire U.S.S.R. [12] Fuel derived from shale oil is utilized in manufacturing enterprises, electric generating stations, on railroads, and in the domestic economy of Estonia. Some form of Estonian shale is used as a fuel in enterprises in the Riga, Vilnyus, Kaunas, Pskov, and Leningrad Oblasts of neighboring Republics.

Over 86 per cent of the 940.9×10^6 kilowatt hours of electricity generated in Estonia in 1955 were based on fuel derived from oil shale. [13] In this same year, Estonia generated 818 kilowatt hours of electricity per person, with shale accounting for the production of 706 of these kilowatt hours. While this is slightly below the national per capita average for energy generated from all fuel resources, it is nevertheless higher than the amount of electricity generated per person from coal alone (see Table 1, Chapter I). If shale were not developed in the Baltic Region, then the geographic distribution of coal consumption would assume new but costly regional patterns.

Radulov, in computing the comparative costs for identical units of energy from various fuel sources, claims that Estonian shale, utilized in the neighboring Leningrad Oblast, is cheaper by 25 to 30 per cent than Donets anthracite and half as expensive as Pechora coal. [14] Computations of this nature do not comprise effective cost data; they are, however, indicators of the geographical use of local fuels.

A major project, now under construction for the utilization of shale, is the Baltic Regional Electric Station, located at Narva in Estonia. The first stage is scheduled for completion in 1960. At that time it will have a capacity of 300×10^3 kilowatts. In 1963, when the second stage is completed, the capacity will reach 600×10^3 kilowatts. Ultimately its capacity will be expanded to $1{,}200 \times 10^3$ kilowatts, with the annual production potential being 3.5×10^9 kilowatt hours, based on a shale consumption of 5×10^6 tons per year. Its full operating capacity will exceed that of the Kakhovka Hydroelectric Station by four times.

References to the Baltic Shale Basin as the "Estonian Donbas" justly serve in emphasizing the importance of this fuel to the region.

The Volga Region.—The Volga has the longest continuous record for shale production in the U.S.S.R. Until the eve of World War II, the output from this region exceeded the Leningrad section of the Baltic Basin. Factual data on the actual position of shale in the energy economy of the Volga Region are unavailable. Some inference of its relative position may be achieved by a knowledge of its development and the nature of its use.

Prior to World War II, and the discovery of oil in the Volga Region, shale was considered an adequate substitute for imported Donets coal in thermoelectric stations. [15] The use of shale was judged economically sound, because its high ash content served as a raw material for high grade cement; and its use curtailed the importance of long haul oil as well as coal.

Today two major thermoelectric generating stations based on oil shale are in operation in the Volga Region. They are located at Syzran and Saratov. [16] There is no evidence, however, that complex refining operations exist at the present time. Experimental work on oil shale of the Obshch-Syrt area reveals that 1×10^6 tons of shale would yield 150×10^6 cubic meters of high calorie gas, 40×10^6 tons of kerosene and benzine, and enough semicoke to operate an electric generating station with a capacity of 25×10^3 kilowatts for one year. [17]

Production Trends 1928-1956.—Shale production in the Soviet Union began in the late 1920's (see Table 10). Output was meager; no more than 600 tons were produced in 1928. These pilot operations were confined to the Volga Region, but by 1932, both the Baltic and Volga Regions were in production, with the major share centered in the Volga area. Warfare interrupted production in the Baltic after 1940, but it continued and increased slowly throughout the war and post war years in the Volga.

Acquisition of Estonia during World War II paved the way for the tremendous expansion which occurred in the shale industry after 1945. Between 1945 and 1956, shale production increased 733 per cent. Most of this increase took place in the Estonian section of the Baltic Basin. Oil shale production in areas under Soviet authority prior to the war increased by just 215 per cent during this same period. In 1956, a scant 35 per cent of the oil shale was mined within territory possessed by the Soviet Government before World War II.

Volga shale production showed a decline of nearly 100 thousand tons between 1955 and 1956. This decline was the result of the introduction of either oil and gas or hydroelectricity into the energy economy of the region.

Kenderal, the only major oil shale deposit in Kazakhstan, has a short and minor history of production. Production began prior to 1940 and continued until 1955; it ceased altogether in 1956. [18]

It can be concluded that production trends give proof to Soviet claims that oil shale is being developed only as a local fuel.

The Regional Pattern and Significance of Shale Production in the National Economy.—Soviet Russia produced 11.5×10^6 tons of oil shale in 1956, an amount theoretically equivalent to 2.3×10^6 tons of

TABLE 9

PRODUCTION OF OIL SHALE IN THE U.S.S.R., SELECTED YEARS, 1928 – 1959
METRIC TONS 10^3

Region and Deposit	1928	1932	1937	1940	1945	1950	1955	1956	1958	1959
Total U.S.S.R.	0.6	318.2	515.0	1,682.0	1,387.1	4,716.0	10,793.0	11,533.0	13,180.0[f]	13,682.0[g]
Baltic Basin	--	73.0	254.0	1,350.5	861.0	3,890.3	9,077.9	9,932.0	---	
Estonia[a]										
Estonian Shale Trust	--	---	---	947.0	367.0	2,666.0	5,708.7	6,097.0	8,960.0[f]	9,091.0[g]
Ministry of Shale Chemical Industry[b]	--	---	---	---	494.0	877.0	1,301.0	1,408.0	---	
Gudov-Chudov	--	73.0	254.0	403.5	---	347.3	2,068.2	2,427.0	---	
Russian Platform										
Izhem[c]	--	---	---	0.8	---	---	---	---	---	
Kostroma	--	---	2.0	23.0	---	---	---	---	---	
Tatar-Ulyanovsk and Kuibyshev-Saratov[d]	0.6	245.2	259.0	306.8	514.0	809.1	1,714.3	1,621.0	---	
Kazakh S.S.R. Kenderal[e]	--	---	---	1.8	12.1	16.8	1.0	---	---	

[a] F. Lopp, ed., *Eesti N.S.V. Rahvamajandus, Statistiline Kogumik* (Tallinn: Eesti Riikilk Kirjastus, 1957), p. 56. In Estonian and Russian. For the years 1940-1956. Ts. S.U. (K. G. Ivanova, ed.), *Promyshlennost' S.S.R.* (Moskva: Gosstatizdat, 1957), p. 166. For the years 1928-1937.

[b] Ts. S.U. (G. I. Groshenkov, ed.), *Narodnoe Khozyaistvo R.S.F.S.R., Statistical Handbook.* (Moskva: Gosstatizdat, 1957), p. 14. For the years 1940-1956. K. G. Ivanova, *ibid.*, p. 166; for the years 1928-1937.

[c] K. G. Ivanova, *ibid.*, p. 166.

[d] G. I. Groshenkov, *ibid.*, p. 14; for the years 1940-1956. K. G. Ivanova, *ibid.*, p. 166; for the years 1928-1937.

[e] K. G. Ivanova, *ibid.*, p. 166; for the years 1940-1950.

[f] Ts. S. U. (S. Ya. Genin, ed.) *Narodnoe Khozyaistvo S.S.S.R. V 1958 Gody, Statisticheskiy Ezhegodnik* (Moskva: Gosstatizdat, 1959), p. 213.

[g] Ts. S. U. (S. Ya Genin, ed.) *Narodnoe Khozyaistvo S.S.S.R. V 1959 Gody, Statisticheskiy Ezhegodnik* (Moskva: Gosstatizdat, 1960), p. 191.

crude oil [19] (see Table 9). The amount and its energy content are insignificant in the total energy framework of the U.S.S.R. Regionally, the pattern of production is simple; yet it is this simplicity that makes oil shale production important to the U.S.S.R. As the following listing shows, approximately 86 per cent of the mining takes place in the Baltic Region and the remaining 14 per cent in the Volga Region.

Region	Per Cent of Geological Reserves	Per Cent of Energy Potential Mineable Reserves	Per Cent of Total Production
Baltic	12.8	26.1	85.9
North and East Russian Platform	11.5	30.7	14.1
Kazakhstan	2.6	1.3	--
Kuznets	1.3	1.0	--
Northeast Siberian Platform	71.8	40.9	--
Total	100.0	100.0	100.0

Localized production can be considered a response to other factors in the geographic environment, principally the lack of energy resources in these regions and the market demands for power.

The Soviet Union has developed these low grade local fuels, rather than adding burdens to already overtaxed railroad facilities. The Baltic area is outstanding in terms of its need for fuel, and this need is aggravated by the fact that the Soviet Baltic fleet has first claim on the products of its shale distilleries. [20] In addition to industrial demand, Soviet strategic planning also contributes to the need for oil shale production in this region and strengthens the importance of the production pattern.

Shale output, as the previous listing indicates, is not in proportion to the distribution of the energy potential. A large potential in isolated Siberia is as yet untouched. Reserves such as the Kuznetsk, that possess high energy content fuels, have remained unused. These factors support the Soviet contention that shale is developed only as a local fuel in areas where market demands have created a need.

Gasification and Gaspipelines.—About half of the shale mined in Estonia is consigned to the Kokhtla-Yarve gasification plant for conversion to fuel gas. [21] This republic manufactures 70 per cent of all shale gas in the Soviet Union. [22] Leningrad, well beyond the Estonian border, exercises first claim on the consumption of this gas. A gas pipeline, 240 kilometers long, connecting the Kohktla-Yarve gasification plant with the Slantsakh plant in the R.S.F.S.R. and then Leningrad, was completed in 1948. [23] Gas transmitted to Leningrad through this pipeline effects an annual energy economy of more than three million cubic meters of firewood and 265 thousand tons of liquid fuel. [24] Almost 2.5 million persons in the city of Leningrad are served with Kokhtla-Yarve gas.

Tallinn, the capital of Estonia, also receives gas by pipeline from Kokhtla-Yarve. A second gasification plant which will also serve Tallinn, is being constructed at Akhtme.

Gasification of shale provides a product that may be inexpensively transported through pipelines over extensive distances. This, in effect, extends the boundaries of regions dependent on local fuel, and creates savings for the entire economy. However, a question to be considered is whether this policy places the supplying republic in a colonial status, and deters its own economic development.

Potential and Production: A Correlative Summary.—Soviet oil shale is produced where market demands dictate. Potential and production are more closely integrated than in the case of coal. Over half the kilowatt hour energy potential of oil shale is located in European Russia, in the Baltic, the European North, and the Volga Region. All production occurs here. Vacant and vast Northeast Siberia, with over 40 per cent of the energy potential, lacks both the demand and the facility for the development of its shale deposits.

In the Baltic region, the absence of other energy resources, plus the energy potential and quality of the shales, are geographical factors that give this region its pre-eminence in the industry. This potential is sufficient to last many years, but with current Soviet emphasis on the development of Siberian regions, European oil shale enterprises may soon reach a static position.

COMPARISON BETWEEN SOVIET AND OTHER OIL SHALES

Nations other than the Soviet Union have and do produce oil shale in commercial quantity. During the nineteenth century, Canada, France, the United States, Australia, and Scotland produced oil from shale. [25] With the exception of Scotland, mining and distillation of shale was discontinued in all of these countries. Since and during World War II, production has been resumed in Australia and the United States. However, none of these countries has been able to extract the same amount of oil from shale as the Estonian shales yield.

Scottish mining has never attained the same level of mechanization as that instituted in Estonia; and according to Ayres and Scarlott, research has never been given the proper incentive. [25] Mines are below the 300 foot level and costs high, but government subsidy has enabled the industry to continue operation.

Australia reinstituted oil shale recovery during World War II, and produced 4 million gallons of motor fuel. Manchuria has large modern operations using the open pit method of mining. Production averages about 16 gallons to the ton, which is much less than the 60 gallon average in Estonia.

Oil shale potential in the United States is much larger than that of the Soviet Union, approximately 700 billion short tons. A pilot plant is now in operation in Colorado. Principal problems in mining United States shale are lack of water and disposal of the ash content. Unlike Estonia and Volga shale operations, where the ash is used to make cement and other building materials, Colorado ash must be dumped as waste.

Other than the United States, where shale oil production is in the experimental stage, all shale producing countries have one thing in common, a lack of local oil supplies. This situation does not apply to the Soviet Union; oil shale is produced in that country to alleviate long coal hauls and because the Soviets inherited a large section of this industry as a result of military conquest.

NOTES – CHAPTER III

1. See Table 10.

2. Bakulev, *op. cit.*, p. 94.

3. P. Antropov, *Our Country's Mineral Wealth* (Moskva: Gos. Izd-vo Polit. Litry, 1956), p. 11.

4. S. Baukov, "Regularity of the Material Composition of Oil Shale in the Baltic Oil Shale Basin," *Eesti N.V.S. Teaduste Akadeemia, Geoloogia Instituudi Uurimused*, No. 11 (1958), 65.

5. Report of the 8th Session of the Soviet State Commission for the Electrification of Russia, *Plan for the Electrification of the R.S.F.S.R.* (Moskva: Gosudarstvennoe Izdatelstvo Politicheskoy Literatury, 1955), p. 196. Second Printing.

6. B. V. Tkachenko, *et al.*, *op. cit.*, p. 240.

7. V. Kalamkarov, "Osnovnye Napravleniya V Razvitii Proizovdstva Topliva V Shestom Pyatiletii," *Planovoe Khozyaistvo*, No. 4 (1957), 17.

8. Computed from data in Oil Shale Appendix – Table I and population data in Ts. S. U., S.S.S.R., (S. Ya. Genin, ed.), *Narodnoe Khozyaistvo S.S.S.R., Statisticheskiy Sbornik* (Moskva: Gosstatizdat, 1956), p. 17.

9. V. Yu. Tarmisto, "New Economic Geography of Estonia," *Geogrifiya V Shkole*, No. 3 (1958), p. 11.

10. Shabarova and Tyzhnova, *op. cit.*, pp. 166-167.

11. P. Antropov, *op. cit.*, p. 11.

12. M. I. Rostovtsev and V. Yu. Tarmisto, *Estonskaya S.S.R. Ekonomiko-Geograficheskaya Kharateristika* (Moskva: Geografgiz, 1957), p. 120.

13. Computed from F. Lopp, ed., *Eesti N.S.V. Rahvamajandus, Statistiline Kogumik* (Tallinn: Esti Riiklik Kirjastus, 1957), p. 59.

14. E. F. Radulov, "Razvitie Slantsevoy Promyshlennosti V S.S.S.R.," *Ugol*, No. 11 (1957), 62. See also Benjamin I. Weitz, *op. cit.*, p. 175. In 1936, this author stated that Gudov Shale cost 30 per cent more in Leningrad than Donets Coal. He said that the utilization of by-products made it economically feasible to use.

15. B. I. Weitz, ed., *op. cit.*, p. 179.

16. L. S. Abramov, ed., *Povolzhe Ekonomiko-Geograficheskaya Kharakteristika* (Moskva: Geografgiz, 1957), pp. 292 and 310.

17. G. N. Cherdantseva, ed., *op. cit.*, p. 217.

18. S. S. Balzak, *et al.*, *op. cit.*, p. 227. This author implies that Kenderal shale was to be developed for nonferrous metallurgy in the Altay.

19. Computed from data in Heinrich Hassman, *Oil in the Soviet Union* (Princeton: Princeton University Press, 1953), p. 82, and data in Table 10.

20. Heinrich Hassman, *op. cit.*, p. 84.

21. M. I. Rostovtsev and V. Yu. Tarmisto, *op. cit.*, p. 121.

22. V. Yu. Tarmisto, *op. cit.*, p. 10.

23. V. A. Gorshkov, "Gazosnabzhenie Leningrada," *Gazovaya Promyshlennost*, No. 11 (1957), 28.

24. V. Yu. Tarmisto, *op. cit.*, p. 11.

25. Eugene Ayres and Charles A. Scarlott, *Energy Sources – The Wealth of the World* (New York: McGraw-Hill Book Company, Inc., 1952), p. 70.

Chapter IV

SOVIET OIL: DISTRIBUTION OF RESERVES, ENERGY POTENTIAL, AND PRODUCTION

The geography of Soviet Russia's oil industry has changed radically in the past twenty years. Today's major reserve and producing area, the Ural-Volga Region, was little more than an object of speculation in 1937. In the intervening years, its oil deposits have been surveyed, proven, and industrially exploited.

Soviet officials declare that their country ranks first in reserves of oil, [1] with 54.8 per cent of the world's total. [2] Representatives of United States oil interests, returning from conferences with Soviet oil personnel in 1956, reported current Soviet known oil reserves as 9.2×10^9 tons, [3] or approximately 26 per cent of the total known crude reserves of the world. [4] This represents an increase of 820 per cent over the known reserves set forth by Gubkin at the XVII International Geological Congress in 1937. [5]

Known oil reserves in the U.S.S.R. have an energy potential of 106.9×10^{12} kilowatt hours. This is less than the energy potential for mineable reserves of oil shale. By way of contrast, the recently discovered coal deposits in South Yakutia have an energy potential three times greater than all of the Soviet known oil reserves. Oil is the most versatile and necessary of fuels in the modern Soviet industrial economy; yet, when compared with other energy resources, its ultimate potential is minute. Were the Soviets to discover no more oil, at the present rate of output, their known reserves would not last for more than 130 years. However, it is safe to predict that new discoveries will take place, probably within the sedimentary basins of Siberia, and that production will grow.

Within the territory encompassed by the present boundaries of the Soviet Union, oil production has increased by 669 per cent since 1913. Production, however, between 1937 and 1955 grew but 152 per cent.

Thus the rate of discovery during this period was greater than the increase in the rate of production. More significant than the overall increase in production has been the shift from the Baku region to the new fields. Geographical changes in the production pattern reflect the changes that have taken place in the distribution of known reserves. These changes are indicative of the increasing significance of oil in the Soviet economy.

SIGNIFICANCE OF OIL IN THE NATIONAL ECONOMY

Oil, preceded only by coal, is currently the second most important fuel in use in the Soviet Union. This position is a post 1950 development, dependent upon recently discovered and exploited oil fields in the Ural-Volga Region.

The significance of oil to the Soviet economy is revealed in two officially computed fuel balances; one depicts the relative amounts of energy resources produced and received, and potentially available as fuel for the national economy; the other shows the relative amounts of the various fuels consumed in the national economy. [6] Both are measured in units of conventional fuel, i.e., bituminous equivalents. (see footnote 1, Table 10, p. 102.) Quantitative comparisons on a thermal basis are thus possible for various fuels.

The Fuel Balance of Received Energy Resources.—Oil has occupied second place in the published reports of fuel balances of received energy resources since 1913 (see Table 10, Part A). Although the secondary position of oil has remained constant, oil did show a relative decline through 1950; a rise was not effected until 1955. This decline can be attributed, in the early years, to Soviet inexperience in oil operations, and in the latter years, to the effects of World War II on the oil industry, and Soviet emphasis on the production of local fuels such as peat and brown coal. In retrospect, the thermal value of oil has consistently placed it in a secondary position in the received fuel balance. Projected developments, however, place oil and coal in a position of near equality by 1970.

The Fuel Balance of Consumed Energy Resources.—In the fuel balance of consumed energy resources, oil currently ranks in second place (see Table 10, Part B), a position it did not attain until after 1950. Prior to that date, wood fuel occupied second place in the consumed fuel balance. The present prominence of oil coincides with intensive oil field development in the Ural-Volga Region. Geographically, this region is more centrally located in relation to the various industrial centers of the Soviet Union than is the Baku Region.

Less than half of the annual oil production of the Soviet Union is

TABLE 10

THE FUEL BALANCES OF THE U.S.S.R.: PART A, ENERGY RESOURCES RECEIVED AND POTENTIALLY AVAILABLE AS FUEL; PART B, ENERGY RESOURCES CONSUMED AS FUEL. SELECTED YEARS 1913-1958, AND PROJECTED YEARS 1965-1970-1975 (IN PER CENT OF CONVENTIONAL FUEL)[1]

Part A. Fuel Balance of Received Energy Resources

Type of Fuel	1913[a]	1927[a]	1932[b]	1937[b]	1940[b]	1950[b]	1955[c]	1958[d]	Projected[d] 1970-1975
Coal	60.7	58.0	59.7	67.1	70.1	73.2	68.5	64.8	36.1
Oil	24.2	20.9	32.2	25.0	21.7	18.9	23.7	21.1	34.6
Natural Gas	--	xxx	1.3	1.6	1.9	2.2	2.6	2.3	23.6
Peat	1.7	4.3	6.7	6.2	6.0	5.0	4.6	4.3	2.5
Oil Shale	--	--	0.1	0.1	0.3	0.7	0.6	0.7	0.8
Wood	13.4	16.8	xxx	xxx	xxxx	xxx	xxx	6.8	2.4

Part B. Fuel Balance of Consumed Energy Resources

Type of Fuel	1913[e]	1932[e]	1937[e]	1940[e]	1950[e]	1953[e]	1954[b]	1955[e]	1958[e]	Proj.[c] 1965
Coal	54.7	59.4	69.5	69.4	75.9	77.5	77.8	77.0	73.9	49.2
Oil	14.1	17.0	11.0	8.3	7.1	8.2	8.4	9.2	10.3	17.8
Natural Gas	--	xxx	xxx	1.9	2.4	2.4	2.3	2.2	6.5	24.8
Peat	1.0	3.7	5.8	6.0	5.0	4.7	4.6	4.3	4.1	3.6
Oil Shale	--	--	0.1	0.3	0.5	0.6	0.6	0.7	0.8	1.1
Wood	30.2	19.9	13.5	14.1	9.1	6.6	6.3	6.6	4.4	3.5

[1] Conventional fuel in the Soviet Union is similar to the bituminous equivalents used in publications of the Western world. Simply stated, all energy resources are converted to units of 7,000 calories; this is equal in heating capacity to a kilogram of Kuznetsk bituminous coal. The method of computation and conversion factors are set forth in: A. A. Rodshtein, *Statistika Energetiki V Promyshlennosti* (Moskva: Gosstatizdat, 1956), pp. 61-63.

[a] Benjamin I. Weitz, *Electric Power Development in the U.S.S.R.* (Moscow: INRA Publishing Society, 1936), p. 157.

[b] G. D. Bakulev, *Voprosy Ekonomiki Topliva V S.S.S.R.* (Moskva: Gospolitizdat, 1957), pp. 18, 30, and 31.

[c] D. Notkin, "Perestroika Toplivnogo Balansa," *Planovoe Khozyaistvo*, No. 1 (1949), 43 and 48.

[d] N. Lavrov, "Novoe V Toplivnom Balanse Strany," *Pravda*, No. 248 (14642), 5 September, 1958, p. 4.

[e] A. G. Burenstam, "Otrazhenie Nekotorykh Voprosov Semiltnego Plana Razvitiya Voprosov Semitelnego Plana Razvitiya Narodnogo Khozyaistvo S.S.S.R., (1959-1965) V Kurse Geografii V Shkole," *Geografiya V. Shkole*, No. 2 (March-April, 1959), 5.

consumed as fuel. A comparison of Part A with Part B in Table 10 indicates that this was especially true in 1955. However, this fact does not diminish the significance of oil; it emphasizes its dualistic importance as a fuel and raw material for other industries. Petrochemical industries, one of the enterprises designated for expansion, will place even greater demands on crude oil as a raw material in the future. [7]

Although oil may now rank second in the consumed fuel balance, it has achieved this position only in recent years. Wood and peat have always been strong competitors, and since 1955, natural gas has received extensive attention. Projected plans for 1965 relegate oil to third place, following coal and natural gas.

Regional Fuel Balance of Consumed Energy Resources.—In the balance of consumed fuels, regionally significant increases of oil are planned for two major geographical divisions of the U.S.S.R. by 1965 (see Table 11). Between 1958 and 1965, European Russia is scheduled for an increase of 5.6 per cent and the Urals Region for a 9.2 per cent increase.

TABLE 11

PROJECTED CHANGES IN THE STRUCTURE OF THE FUEL BALANCE OF CONSUMED ENERGY RESOURCES BY REGION, 1958 - 1965 (IN PER CENT OF CONVENTIONAL FUEL UNITS)[a]

Type of Fuel	European Russia		Urals Region		Eastern Regions	
	1958	1965	1958	1965	1958	1965
Coal	68.1	49.2	80.8	46.6	86.3	82.2
Oil	10.9	16.5	11.3	20.5	7.8	7.9
Natural Gas	9.6	25.3	0.9	28.1	0.9	6.0
Peat	6.0	4.7	1.7	1.1	--	0.2
Oil Shale	1.3	1.6	--	--	--	--
Wood	4.1	2.7	5.3	3.7	5.0	3.2
Gas from underground gasification	--	--	--	--	--	0.5
Total	100.0	100.0	100.0	100.0	100.0	100.0
Regional Consumption[b]	64.5	59.1	16.7	16.9	18.8	24.0

[a]Adapted from: Yu. Bokserman, V. Kalamkarov, and A. Kortunov, "Zadachi Razvitiya Gazovoy Promyshlennost," *Planovoe Khozyaistvo*, No. 12 (1958), 34.

[b]D. Notkin, "Perestroika Toplivnogo Balansa," *Planovoe Khozyaistvo*, No. 1 (1958), 48.

In 1958, oil occupied second place in the consumed fuel balance of European Russia, supplying 10.9 per cent of the conventional fuel. By 1965, oil is destined to supply 16.5 per cent of the fuel, as measured in conventional fuel units. Even with this increase, oil will be surpassed by gas as the second most important fuel.

A similar situation exists in the Urals, where oil will increase from 11.3 per cent to 20.5 per cent in the balance of consumed fuels. Here too it will be displaced from a secondary to a tertiary position by the use of gas.

Lower production costs per ton for oil, and a higher energy content indicate its use and substitution for coal wherever feasible. In the Urals, this is particularly true, for a ton of Bashkir coal requires eight times more capital investment than does a ton of local crude oil. [8] According to Notkin, oil shale, a major local fuel in the Kuibyshev oblast, costs ten times more per ton (in conventional fuel units) to produce, than does local crude oil. Such drastic changes in the regional fuel balance of the Soviet economy can only be effected because of a definitive understanding of known oil reserves.

OIL RESERVES - EVALUATION OF AN ELUSIVE FIGURE

Techniques involved in evaluating oil reserves are as complex as the geology of oil itself. Theories of oil migration and accumulation in the sedimentary strata of the earth's crust are constantly under intense geological and geophysical investigation. Until these theories are confirmed and other related geological problems are solved, reasonably exact estimates of oil reserves will be difficult. The knowledge that even the most reliably estimated reserves are, in part, subjective is imperative to an understanding of Soviet oil claims.

Extensive survey activity, accelerated by official decree, is destined to result in an increase of known reserves. Over 45 per cent of the territory of the U.S.S.R. contains sedimentary formations capable of yielding oil, and much of it is now being explored. In 1957 alone, the Soviet Union had 980 geophysical survey teams in the field searching for oil, [9] and a total of 2.8×10^6 meters of exploratory drilling took place. [10]

With these facts in mind, it is possible to examine estimates of Soviet oil reserves. Two categories of reserves will be discussed: possible and known.

Possible Reserves.—Possible reserves in a region are derived by computing the volume of sedimentary strata theoretically capable of containing oil and, subsequently, the amount of oil that the strata might conceivably contain. Conjecture is a major factor in this type of computation, because the actual existence of oil may or may not have been established. Despite this, possible reserves may serve as a valuable guide in directing survey crews to promising oil formations (see Table 12).

Known Reserves.—The geography of a nation's oil potential is reflected in the distribution of its known reserves, which are those deposits where the existence of oil has been proven by test drilling and geophysical exploration. [11] Deposits in this category are suitable for production or are actually being produced. The more definitive nature of the known category dictates its use in this study.

In 1955, the Soviets reported known oil reserves as 9.2×10^9 tons (see Table 12), an increase of 820 per cent over the widely accepted 1937 estimate. When compared with the 426 per cent increase in geological reserves for coal during this same period, the gain for oil appears spectacular. Emphasis must again be placed on the cautious nature of earlier Soviet estimates and the limited knowledge of the

TABLE 12

CHRONOLOGICAL LISTING OF ESTIMATES OF POSSIBLE RESERVES AND KNOWN RESERVES OF SOVIET OIL COMPUTED BETWEEN 1937 AND 1955, METRIC TONS 10^9

Year	Possible Reserves	Known Reserves	Authority and Source
1937	---	0.819	Department of State, *Energy Resources of the World* (Washington: G.P.O., 1949), p. 71.
1937	6.4	0.882	I. M. Goubkin,[b] "World Petroleum Reserves," *Report of the XVII Session, International Congress* (Moscow: 1939) p. 184.
1938	8.5	---	I. M. Gubkin,[b] quoted by Heinrich Hassman, *Oil in the Soviet Union, History, Geography, Problems* (Princeton: Princeton University Press, 1956), p. 65.
1939	8.7	---	State Planning Commission, *Third Five-Year Plan for the Development of the National Economy of the U.S.S.R., 1938-1942* (Moscow: Gosplanizdat, 1939), p. 170.
1948	27.8[a]	---	F. Julius Fohs, "Petroliferrous Provinces of the Union of Soviet Socialists Republics," *Bulletin American Association of Petroleum Geologists*, Vol. 32, No. 3 (1948).
1950	24.0[a]	---	Eugene Stebinger, in *World Geography of Petroleum* (Princeton: American Geographical Society – Princeton University Press, 1950), p. 238.
1950	---	4.6[a]	Dr. J. Brian Eby, "How Oil Production in Russia Has Gained Since World War II," *World Oil*, Vol. 143, No. 3 (1956), 182.
1952	---	4.5[a]	*Petroleum Press Service*, March 1952, pp. 97-98.
1953	---	6.7[a]	Dr. J. Brian Eby, op. cit., p. 182.
1955	---	9.2[a]	Dr. J. Brian Eby, op. cit., p. 182.

[a]Converted from barrels to metric tons in accordance with standards set forth in – Department of State, *Energy Resources of the World* (Washington: G.P.O., 1949), p. 124.

[b]In publications of the XVII Session of the International Geological Congress, this name is spelled Goubkin. The correct transliteration found in all subsequent geological publications is Gubkin.

country's geological status at the time. Actual oil production in the U.S.S.R. between 1937 and 1957, excluding the war years, totaled 859×10^6 tons, an amount nearly equal to the 882×10^6 tons estimated for known reserves in 1937. [12]

Most of the growth in known reserves was achieved after 1950. In the 13-year period from 1937 to 1950, known reserves increased from 882×10^6 tons to 4.6×10^9 tons. Between 1950 and 1955, an additional 4.6×10^9 tons were added to the stock. The 100 per cent increase effected in this five year period is not without parallel in the Western World. A 208 per cent increase occurred in the Free World between 1947 and 1957, when known reserves of crude oil rose from 9.5×10^9 tons to 29.5×10^9 tons. [13]

New regional patterns in the geography of Soviet Russia's oil industry have evolved from the increase in known reserves. Col. Miksche has shown that the defense citadel of the Soviet Union occupied the triangle Moscow to Omsk to Baku. [14] The vulnerability of Baku undoubtedly influenced the Soviet change in attitude which resulted in greater survey activity in the more strategic Ural-Volga Region. As late as 1940, over 70 per cent of all exploratory drilling was conducted in the Transcaucasus and the North Caucasus Region, but by 1955 just 25 per cent of the exploratory drilling occurred there. [15]

As a result of intensive surveying in the Ural-Volga Region, Soviet Russia's known oil reserves have increased in the past decade. Strategic considerations provided the impetus in the search for oil outside the Caucasus Region. Discovery of large deposits of oil in the "Second Baku" or Ural-Volga Region confirmed, in part, earlier Soviet estimates of possible reserves for that area. Because of these oil deposits, a new energy base for the country has been developed.

THE KILOWATT HOUR ENERGY POTENTIAL OF SOVIET OIL RESERVES

The Total Energy Content of Soviet Oil Reserves.—Soviet Russia's known oil reserves contain a potential 106×10^{12} kilowatt hours (see Table 13). Thus, the total per capita energy potential would be 534×10^{3} kilowatt hours per person. Coal and oil shale have a greater energy potential than does oil, as the listing illustrates:

Type of Energy Resource	Total Energy Potential Kwt. Hrs. 10^{12}	Per Capita Energy Potential Kwt. Hrs. 10^{3}
All coal	43,233.1	241,000.
Oil Shale	120.5	602.
Oil	106.9	534.

The energy potential of known oil reserves is small in comparison with coal. It does not quite equal the energy potential of the coal deposits on Sakhalin Island and has but a third of the energy potential contained in the newly discovered coal basin in South Yakutia.

Nevertheless, in spite of its small share in the total potential energy framework, oil achieves prominence because of its high thermal value per unit of weight, its low extraction cost in comparison with other energy resources, and the relative ease with which it can be transported. The current importance of oil as an energy resource is heightened by the favorable geographic distribution of newly established known reserves.

TABLE 13

THE DISTRIBUTION OF SOVIET KNOWN OIL RESERVES AND THEIR KILOWATT HOUR ENERGY POTENTIAL

Region	Known Reserves Metric Tons 10^6	Energy Content Kwt. Hrs. 10^6	Per Cent Distribution
Ural-Volga	7,424.[a]	86,326,272	80.70
Southwestern (including)			
Azerbaidzhan	994.[b]	11,558,232	10.80
North Caucasus	383.[c]	4,453,524	4.17
Georgia	3.[d]	34,884	0.03
Ukraine	6.[e]	69,768	0.06
Komi-Ukhta	13.[f]	151,164	0.14
Sakhalin	40.[g]	465,120	0.44
Kazakh-Emba	110.[h]	1,279,080	1.19
Middle Asia (including)			
Turkmen S.S.R.	200.[i]	2,325,600	2.17
Uzbek S.S.R.	20.[j]	232,560	0.22
Kirgiz S.S.R.	5.[j]	58,140	0.06
Tadzhik S.S.R.	2.[j]	23,256	0.02
Total	9,200	106,977,600	100.00

[a]Data on total reserves tonnage contained in: Dr. J. Brian Eby, "How Oil Production in Russia Has Gained Since World War II," *World Oil*, Vol. 143, No. 3 (August, 1956), 182, were utilized for computing and estimating the distribution of known reserves in accordance with percentages for regional distribution of reserves cited for individual regions below.

[b]Computed from percentages given for the distribution of known reserves in the Ural-Volga Region in A. A. Keller, *Neftyanaya i Gazovaya Promyshlennost S.S.S.R. V Poslevolnnye Gody* (Moskva: Gostoptekhizdat, 1958), p. 7 and M. Brenner, "Problems of Petroleum in the Perspective of Development of the U.S.S.R. Economy," *Problems of Economics*, Vol. 1, No. 4 (August, 1958), 13. Translated contents of *Voprosy Ekonomiki*, Feb., 1958, by International Arts and Sciences Press.

[c]Computed from percentages given for the known reserves in A. A. Keller, *op. cit.*, p. 7, and V. Kalamkarov, "Osnovnye Nopravleniya V Razvitii Proizovdstva Topliva V Shestom Pyatiletii," *Planovoe Khozyaistvo*, No. 4 (1957), 19.

[d]Computed from percentages given in Keller, *op. cit.*, p. 7; Kalamkarov, *op. cit.*, p. 19; and N. A. Kalinin, "40 Let Poiskov i Razvedki Neftvanykh i Gazovykii Mestrozhdeniy," *Razvedka i Okhrana Nedr*, No. 11 (1957), 37-42.

[e]Estimated on the basis of S. F. Fedorov, *Neftyannyye Mestroozhdeniya Sovietskogo Soyuza* (Moskva: Gostoptekhizdat, 1939); I. M. Goubkin, "World Petroleum Reserves," *Report of the XVII Session, International Geological Congress, Vol. I* (Moscow: 1939), pp. 182-184; N. A. Kudriavtzev, "The Oil Fields of Georgia," *Abstract of Papers, International Geological Congress, XVII Session* (Moscow: 1937), pp. 10-11; A. J. Krems, *The Petroleum Excursion, The Georgian S.S.R.* (Moscow: 1937), 65 pp; and production trends in Ts. S.U., S.S.S.R. Statisticheskoe Upravlenie Gruzinskoy S.S.R. (S. M. Pirumov, ed.), *Narodnoe Khozyaistvo Gruzinskoy S.S.R., Statisticheskii Sbornik* (Tbilisi: Gosstatizdat, 1957), p. 42.

[f]Estimated from data in: T. T. Gonta, N. A. Gorev, I. F. Klitochenko and K. F. Mikhailov, *Neft i Prirodnyy Gaz Ukrainy* (Moskva: Gostoptekhizdat, 1957), 79 pp; Heinrich Hassman, *Oil in the Soviet Union, History, Geography, Problems* (Princeton: Princeton University Press, 1953), pp. 77-82; Akademiya Nauk Ukrainskoy S.S.R., Institut Ekonomiki, *Ukrainskaya S.S.R., Chast I* (Moskva: Geografgiz, 1957), pp. 56-57, 226-227; Akademiya Nauk Ukrainskoy S.S.R., Institute Ekonomiki, *Ukrainskaya S.S.R. Chast II*, (Moskva: Geografgiz, 1958), pp. 298-299; and production trends in: Tsentralne Statistichne Upravlinnya pri Radi Ministriv S.R.S.R., Statistichne Upravlennya Ukrainskoi R.S.R., (P. S. Bulgokov, ed.), *Narodne Gospodarstvo Ukrainskoi R.S.R., Statistichniy Zbirnik* (Kiev: Derzhstatvidav, 1957), p. 40.

DISTRIBUTION OF RESERVES AND THEIR KILOWATT HOUR ENERGY POTENTIAL

Known oil reserves in the Soviet Union are distributed throughout 11 major geographical regions, extending from the western border of the Ukraine to the Far Eastern island of Sakhalin (see Fig. 12). On the basis of total tonnages given for oil by Dr. J. Brian Eby, and published percentages for the distribution of known reserves by various Soviet authors, [16] it has been possible to compute definitively the regional distribution for 95 per cent of the known Soviet oil reserves. Estimates for the amount and distribution of the remaining 5 per cent of the known reserves are based on earlier geological reports, primarily Gubkin and Fedorov, [17] and recent geological surveys, plus survey ratios required in relation to annual production. Sources used in arriving at estimates are cited as footnotes to Table 14. However, the majority of the material consulted appears in this chapter's bibliography.

Today over 95 per cent of these reserves with their proven energy potential are located in European Russia, including the western slope of the Urals. One oil region alone, the Ural-Volga, often called the "Second Baku," contains 80 per cent of the total 9.2×10^9

---

[g]Estimated from data in: S. F. Fedorov, op. cit.; I. M. Goubkin, *op. cit.*; N. M. Tikhonovitch, "The Oil Fields of the Ukhta-Pechora Region," *Abstract of Papers, International Geological Congress, XVII Session* (Moscow: 1937), pp. 21-22; Heinrich Hassmann, *op. cit.*, pp. 97-99.

[h]Estimated from I. M. Goubkin, *op. cit.*, pp. 182-184; M. G. Tanassevitch, "The Oil Fields of North Sakhalin," *Abstract of Papers, International Geological Congress, XVII Session* (Moscow: 1937), pp. 24-25; V. G. Udovenko, *Dolniy Vostok, Ekonomiko-Geograficheskaya Kharakteristika* (Moskva: Geografgiz, 1957), pp. 19, 196-197; and production trends in Heinrich Hassman, *op. cit.*, p. 103.

[i]Estimated from S. F. Fedorov, *op. cit.*; I. M. Goubkin, *op. cit.*, p. 1; S. V. Shumilin, "The Oil Fields of the Emba Region," *Abstract of Papers, International Geological Congress, XVII Session* (Moscow: 1937), p. 16; I. G. Permiakov, "The Emba Salt Domes," *Abstract of Papers, International Geological Congress, XVII Session* (Moscow: 1937), p. 17; N. N. Pagov, *Kazakhstan* (Moskva: Geografgiz, 1953), pp. 121-132. V. Ya. Avrov, "O Protsessakh Neftenakopleniya Solyanokupolnykh Strukturakh Prikaspiiskoy Depressii," *Doklady Akademii Nauk S.S.S.R., Noyaya Seriya* No. 4 (1951); and Akademiya Nauk S.S.S.R., Institut Geografii Akademiya Nauk Kazakhskoy S.S.R. Sektor Geografii, *Kazokhskaya S.S.R.* (Moskva: Geografgiz, 1957), pp. 77, 688-689.

[j]Estimated from data in: I. M. Goubkin, *op. cit.*; V. I. Kulikov, "Oil Fields on the Turkmen S.S.R.," *Abstract of Papers, International Geological Congress, XVII Session* (Moscow, 1937), p. 11; Z. G. Freikin, *The Turkmen S.S.R.* (Moskva: Geografgiz, 1954), pp. 109-222-223-228-229; and production trends in Z. G. Freikin, *Turkmenskaya S.S.R., Ekonomiko-Geograficheskaya Kharakteristika* (Moskva: Geografgiz, 1957), p. 184-185; and Ts. S. U. S.S.S.R., Statisticheskoe Upravlenie Turkmenskoi S.S.R., A. Charnyev, ed., *Norodnoe Khozyaistvo Turkmenskoy S.S.R., Statisticheskiy Sbornik* (Ashkhabad: Gosstatizdat, 1957), p. 27.

[k]Estimated from data in: *Neftyannoye Khozyaistvo*, No. 4 (1948) 35-41; I. M. Goubkin op. cit. 184; K. P. Kalitzky, "The Oil Fields of the Central Asian Republics," *Abstract of Papers, International Geological Congress, XVII Session* (Moscow, 1937), pp. 11-12.

tons of known reserves (see Table 13). Two other regions, the Azerbaidzhan-Baku and the North Caucasus possess an additional 15 per cent. The remaining 5 per cent of the reserves are dispersed over widely separated regions.

Middle Asian deposits account for approximately 3 per cent of the reserves. Western Turkmenistan is the primary oil area, with 2.72 per cent of the Middle Asian reserves. The balance, 0.30 per cent, is located in the Fergana Valley and the Tadzhik Depression.

Kazakhstan, with its famed oil producing region at Emba, has 1.3 per cent of the total known reserves, while Sakhalin Island in the Far East contains 0.44 per cent. Ukhta and environs in the Komi A.S.S.R. of the North have 0.15 per cent. The Ukraine in the South has the remaining 0.06 per cent.

The Ural-Volga Region.—In slightly over a decade, the Ural-Volga Region has surpassed the famed Baku Region to become the Soviet's major oil reserve base. In 1956, Soviet oil authorities credited this region with exactly 80.7 per cent of the total known reserves. [18] Distributed throughout are 7.4×10^9 tons of oil with an energy potential of 86.3×10^{12} kilowatt hours. Geographic location imparts a special significance to the energy potential which is located astride all major communication lines between the Central Industrial Region and the Urals. It extends into the Udmurt A.S.S.R. in the North and to the Stalingrad Region in the South. The western slope of the Urals forms its eastern boundary, and in the West it extends into the Penza and Ulyanovsk oblasts of the Central Regions.

Most of European Russia, including the Urals, can be supplied with oil from its great potential. The Baltic States, Belorussia, the West and Northwest are some distance from this major source of oil, but they are recipients. In fact, oil from the Ural-Volga deposits not only supplies these regions, but is shipped beyond to Czechoslovakia, Yugoslavia, Poland, and other countries. [19] Someday pipelines from the "Second Baku" may pour oil into energy-poor Western Europe.

Tectonic boundaries for this vast region are under study. The Caspian Depression in the South and the granitic mass of the Urals in the East form the known boundaries. [20] Surveys have revealed the existence of oil to the west and north in the areas mentioned above, but further work must be completed before definite limits are set in these directions. Oil may yet be discovered at the gates of Moscow.

Equal in size to Spain, the Ural-Volga Region is not underlain by one continuous strata of oil; rather, oil is found in many widely scattered deposits. Only by the most advanced survey techniques and study of the formation and migration of oil have the Soviets been able to discover new deposits of oil in this field. [21]

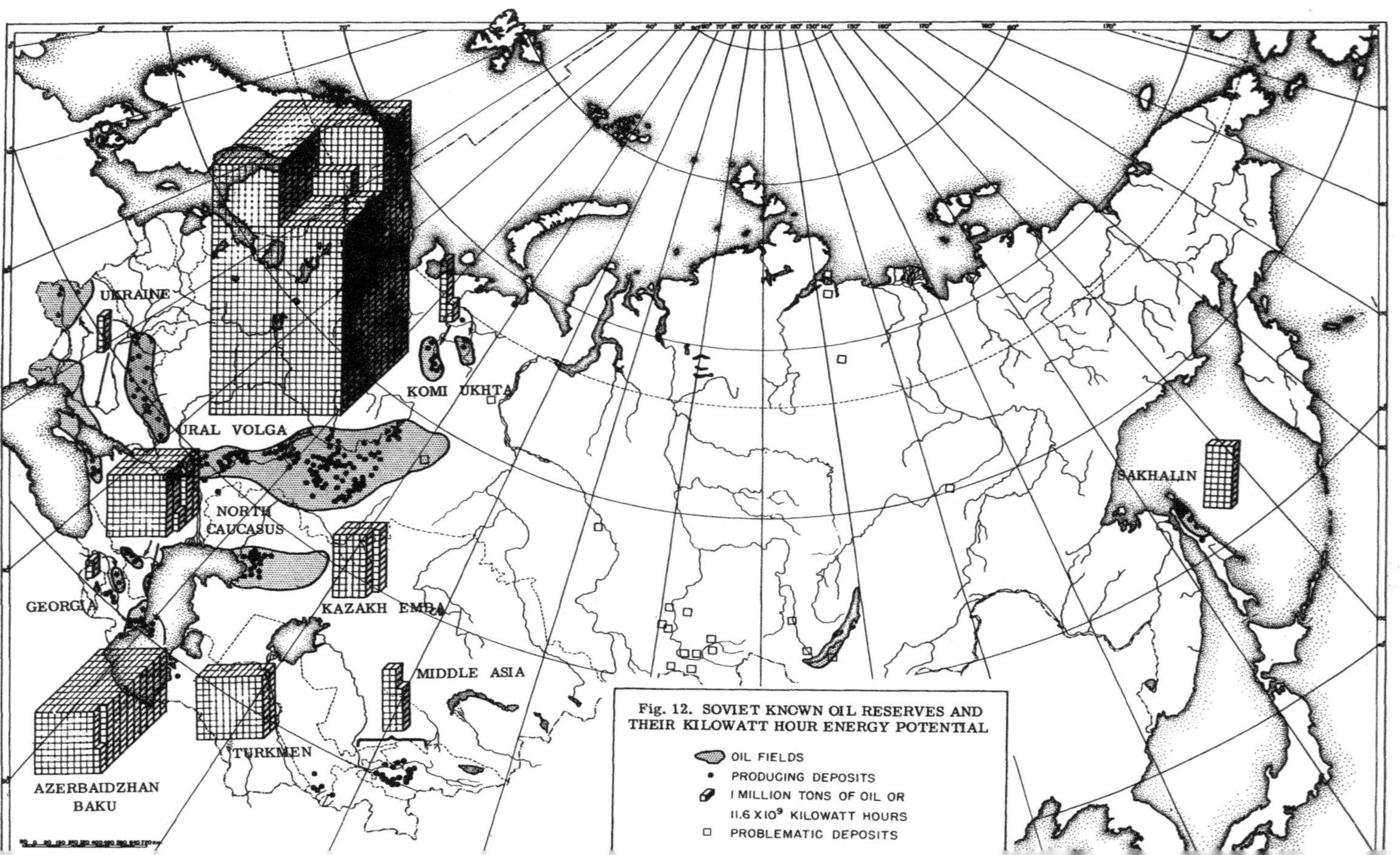

Fig. 12. SOVIET KNOWN OIL RESERVES AND THEIR KILOWATT HOUR ENERGY POTENTIAL

In 1940, prior to the war, not more than 15.8 per cent of the known reserves and their kilowatt hour energy potential were located here, but by the end of World War II, in 1946, this region possessed 30.3 per cent of the total oil energy potential of the country. The increase to 80.7 per cent of the nation's total in 1956 is a reflection of the progressively increasing survey activity (see Fig. 13) which was stimulated by invasion and perpetuated by success. M. Brenner leaves no doubts concerning the success of this survey drilling when he explains that the growth in reserves per drill hole is ten times greater in the Ural-Volga Region than it is in the Caucasus or Middle Asia. [22]

Between 1940 and 1956, survey drilling increased from 29.4 per cent of the nation's total to 67.4 per cent. [23] During the war years, 1941-1945, the Soviets completed a total of 427,000 meters of survey drilling in the Ural-Volga Region. In the immediate post war period, 1946-1950, they drilled $1{,}587 \times 10^3$ meters, and between 1951 and 1955 survey drilling totaled $4{,}298 \times 10^3$ meters. [24]

In place of specific published data, the key to distribution of reserves in this extensive region is provided by a study of survey drilling and reports of new discoveries. This material is reported by political unit (see Table 14). In the fourth Five Year Plan, 55.4 per cent of the survey drilling was conducted in the Bashkir and Tatar Autonomous Republics and the Kuibyshev Oblast. At that time 75 per cent of the newly discovered oil deposits were located in these political units, and 60 per cent of the operational drilling was instituted there. In the fifth Five Year Plan these units had approximately 57 per cent of the survey drilling, 59 per cent of the new discoveries, and 72 per cent of the operational drilling. These facts, plus the data contained on the map in Figure 12, would indicate that oil reserves in the Ural-Volga Region are located primarily in these three political units. Reportedly, the two largest deposits are the Tuimazy, in the Bashkir Republic, and the Romashkin, in the Tatar Republic. [25]

Saratov and Stalingrad Oblasts combined have had an average of slightly more than 10 per cent of the survey and operational drilling since 1946. These oblasts had 21 per cent of the total Ural-Volga discoveries in the period 1951-1955. A large number of these deposits are highly valued by the Soviets for their gas content.

Survey drilling in the other areas of the Ural-Volga Region can be considered as logical attempts to extend the northern and western boundaries of this oil province.

Azerbaidzhan-Baku.—Azerbaidzhan, once the oil capital of the world, now is reported to contain just 10.8 per cent of Soviet known oil reserves, and their kilowatt hour energy potential. The 994×10^6 tons of oil located in this region have an energy potential of 11.5×10^{12} kilowatt hours. Reserves have increased by more than 80 per cent

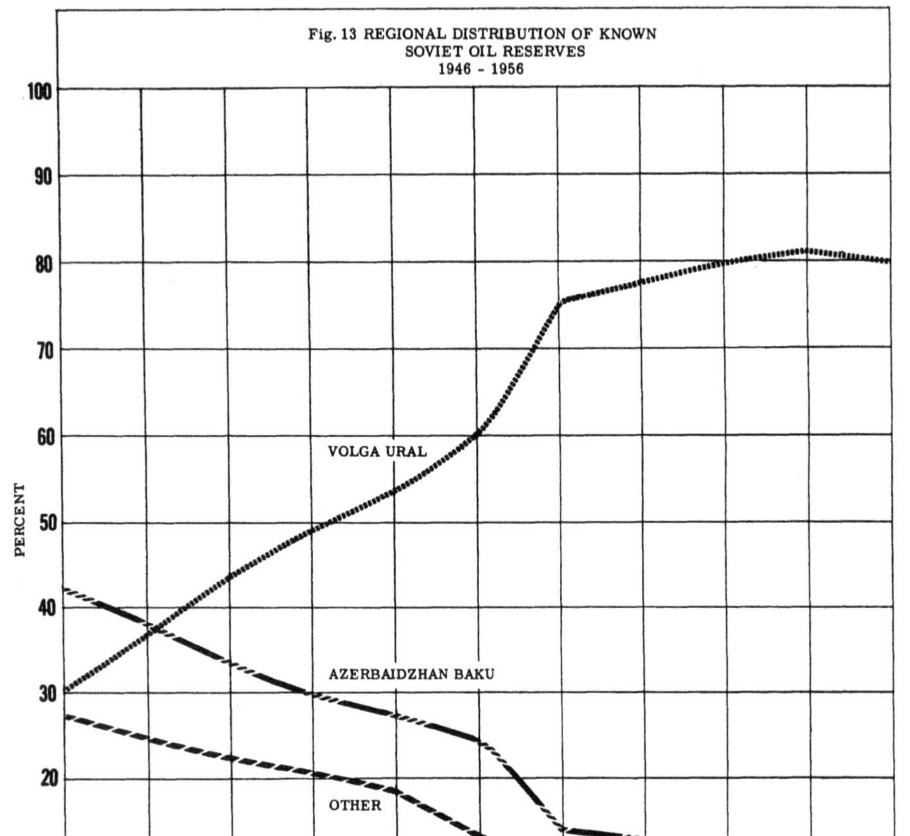
Fig. 13 REGIONAL DISTRIBUTION OF KNOWN
SOVIET OIL RESERVES
1946 - 1956
100
90
80
70
60
50
40
30
20
10
PERCENT
VOLGA URAL
AZERBAIDZHAN BAKU
OTHER
1946
1947
1948
1949
1950
1951
1952
1953
1954
1955
1956

TABLE 14

SURVEY AND OPERATIONAL DRILLING, AND THE DISCOVERY OF NEW DEPOSITS AND POOLS OF OIL AND GAS IN THE URAL-VOLGA REGION DURING THE FOURTH AND FIFTH FIVE YEAR PLANS

Region	Survey Drilling[a] 1946-1950 Meters 10^3	Per Cent	New[b] Discoveries 1946-1950	Operational Drilling[a] 1946-1950 Meters 10^3	Per Cent
Total Ural-Volga	1,724.1	100.0	36/89	2,850.1	100.0
Bashkir A.S.S.R.	504.1	29.2	9/9	906.1	31.2
Tatar A.S.S.R.	112.0	6.5	4/6	169.1	5.9
Perm Oblast	103.9	6.0	1/3	306.4	10.8
Kuibyshev Oblast	339.9	19.7	14/19	667.4	23.1
Orenburg Oblast	38.3	2.2	3/6	183.9	6.5
Saratov and Stalingrad Oblast	240.0	13.9	0/14	331.0	11.6
Udmurt A.S.S.R., Penza, Ulyanovsk	385.9	22.5	---	285.2	10.9
		1951-1955			
Total Ural-Volga	4,296.0	100.0	71/156	8,454.6	100.0
Bashkir A.S.S.R.	1,332.8	31.0	27/31	2,723.9	32.2
Tatar A.S.S.R.	339.5	9.4	8/16	1,838.8	21.8
Perm Oblast	275.3	6.4	5/12	406.8	4.8
Kuibyshev Oblast	696.6	16.2	8/32	1,495.8	17.7
Orenburg Oblast	56.2	1.3	0/8	214.8	2.5
Saratov and Stalingrad Oblast	464.8	10.8	11/28 } 4/21 }	847.3	10.0
Udmurt A.S.S.R., Penza, Ulyanovsk	1,230.7	24.9	8/8	927.2	11.0

[a]Adapted and computed from A. A. Keller, *Neftyanaya i Gazovaya Promyshlennost S.S.S.R. V Poslevolnnye Gody* (Moskva: Gostoptekhizdat, 1958), p. 28.

[b]Adapted from A. A. Trofimuk, *Uralo-Povolzhe-Novaya Neftyanaya Baza S.S.S.R.* (Moskva: Gostoptekhizdat, 1957), p. 141.

over Gubkin's 1937 estimate, but the region has lost its primary position in the Soviet oil industry because of the tremendous discovery rates in the "Second Baku" Region (see Fig. 13).

Unlike the Ural-Volga, with its widely dispersed deposits, the reserves of the Azerbaidzhan-Baku Region are concentrated in a relatively small area, with as many as 22 producing oil horizons. [26] After nearly a century of production, the Soviets have been forced to survey for oil off shore in the Caspian Sea, and at greater depths in the established producing areas. The search for offshore oil is not a recent development in Azerbaidzhan. Kuliev and Safarov claim that the world's first successful well, drilled from the shore out under the sea, was at Bukhty-Ilicha in 1925. [27] Other shore to sea wells, mainly on Atrem Island, followed in the early 1930's. In 1935, the first well actually drilled at sea was off this island. [28] The most successful offshore drilling occurred at Nefti Kamin, a reef off the tip of the Apsheron Peninsula. Drilling was conducted in sea depths of 24 to 30 meters during the fifth Five Year Plan [29] and was extended to depths of 40 meters in the sixth Five Year Plan. [30]

In an effort to locate more oil in the Baku Region, the Soviets have resorted to deep drilling. The deepest well in the country is found in the southeast section of the Apsheron Peninsula. Completed in 1955, this well is 4,812 meters deep. [31]

A third and more costly method of extracting oil is planned for some of the older producing areas around Baku. This is the mining or shaft method, currently utilized at Ukhta and in the Ukraine. Preparations for mining of light oil at the old Balakhani wells near Baku were underway in 1956. [32]

Maintaining and increasing reserves of oil in the Baku region have been accomplished under difficult circumstances. The return of oil per meter of drilling has been declining for some time, [33] with the result that the capital investment per ton of oil produced has increased. [34] As the volume of survey drilling in the Soviet Union grows, the percentage of total drilling allocated to Azerbaidzhan decreases. In the years of the fourth Five Year Plan, 24.2 per cent of all survey drilling occurred in Azerbaidzhan, but in the fifth Five Year Plan, 13.7 per cent of the survey drilling was conducted in this Republic. By 1956, Azerbaidzhan's share of the survey drilling was down to 8.8 per cent of the Union's total. [35]

Known reserves of oil and their kilowatt hour energy potential in the Azerbaidzhan-Baku Region rank second after the Ural-Volga Region. Most of the known reserves today are in deep lying strata, or in offshore deposits, making extraction difficult and expensive.

The Northern Caucasus Region.—Approximately four per cent of the Soviet Union's known oil reserves are located in the North Caucasus Region. Historically, the Northern Caucasus has occupied second place, after Azerbaidzhan, in the regional distribution of oil reserves. With the emergence of the Ural-Volga Region as the primary oil reserve base, North Caucasian deposits have been relegated to a tertiary position.

Known oil reserves in the Northern Caucasus total 383×10^6 metric tons. The energy potential of these reserves exceeds 4.4×10^{12} kilowatt hours. Surveyed reserves are concentrated primarily in the Krasnodar Krae, which in recent years has surpassed the declining Maikop and Grozny Regions. In the past 10 years, reserves in the entire North Caucasus increased 75 per cent, [36] a direct result of discoveries in the Krasnodar Krae. Oil reserves in this section of the North Caucasus increased 282 per cent during the same period. Discovery of the Anastasiev-Troits deposit accounted for a great deal of the growth.

Of the 84.2 thousand square kilometers comprising the territory of Krasnodar Krae, 50 thousand have been evaluated as a prospective oil and gas area. In 1956, 20 per cent of the 50 thousand square kilometers had been surveyed, and known reserves were established in one-third of the area. [37]

Reserves of oil in the Dagestan Autonomous Republic are only partially surveyed at the present time.

The Ukraine.—The Ukraine has an estimated 6×10^6 tons of oil, or .06 per cent of the Soviet known reserves. When measured in terms of their energy potential, these reserves contain 69.7×10^9 kilowatt hours. Brown coal deposits of the Dneper Basin have an energy potential several thousand times greater than all of the Ukrainian oil.

Three oil provinces are found in the Ukraine: the Eastern Carpathians, the Dneper-Donets Region, and the Black Sea-Crimea Region. Most of the known oil reserves are concentrated in the Eastern Carpathian Region. [38] Between 1952-1954, the Soviets completed 670,000 meters of survey drilling in this region, or 628 meters more than was drilled in exploratory work in all of Siberia and the Far East in the fourth Five Year Plan. [39] Relatively little is known concerning the oil capacity of the Dneper-Donets Region. Although oil was discovered here in 1933-1934, production was not instituted until 1952. Survey drilling has produced oil in the Peninsula section of the Black Sea-Crimean Region, but commercial exploitation has not been instituted.

Shaft mining of oil was begun in the Borislav region of the East Carpathians in the first quarter of 1958. Shafts, 6 meters in diameter and approximately 120 meters in depth, with lateral drifts, drilled into the oil bearing strata are being used. Secondary recovery is one reason for this type of operation. Another reason is the high ozocerite content which inhibits the flow of oil. [40]

The Komi-Ukhta Region.—The existence of oil in the Ukhta region of the Komi A.S.S.R. has been recognized since the time of Peter the Great. Known reserves of 13×10^6 tons are estimated. [41] These reserves have an energy potential of 151.1×10^9 kilowatt hours, or .14 per cent of the Soviet total. Quality oil from the Ukhta wells has a high pour point and it freezes at relatively high temperatures. Oil mining, pioneered prior to World War II, was the method adopted by the Soviets to obtain this high pour oil. Leningrad, in the Northwest, is the chief recipient of Ukhta oil. Once considered a distinct oil province, there is evidence today that this region is but a northern extension of the Ural-Volga oil province.

Sakhalin Island.—Sakhalin Island contains the only known industrially exploited oil deposits in Soviet Siberia and the Far East. Known reserves are estimated as 40×10^6 tons, or .44 per cent of the total Soviet oil potential. The energy potential of this Far Eastern oil province is 465.1×10^9 kilowatt hours. Oil was first produced on

Sakhalin Island in October, 1928. [42] At the present time there are over 2,000 producing wells in ten areas.

Geographic location, not the volume or quality of the oil, lies behind the significance of the region. As the only producing oil region in the Far East, it supplies the important naval base of Vladivostok, as well as other cities and industries of the area. In addition to shipping oil from the island in tankers, a pipeline is reported to have been built from Okha, on the island, to Nikolayevsk across the Tatar Strait at the mouth of the Amur, then south through Komsomolsk to Khabarovsk.

Kazakh-Emba Region.—The Emba Region is located in the Caspian Depression along the northeastern shores of the Caspian Sea. Known reserves have an estimated 110×10^6 tons of oil, with an energy content of 1.2×10^{12} kilowatt hours. This desert-steppe region of Kazakhstan possesses 1.19 per cent of Soviet Russia's known oil reserves.

Oil is found in salt dome formations. Palgov stated, in 1953, that not less than 1,500 of these formations were known. [43] Some domes are 10 kilometers in diameter. Earlier, in 1937, Shumilin reported the known existence of just 300 domes. [44] At that time only 20 domes were being exploited and prospected. Not all of the domes, however, contain commercial amounts of oil. Permiskov estimated that not more than 65-70 per cent would contain commercial deposits of oil. [45] Lack of water in this arid region hinders exploitation.

Survey drilling during the fourth and fifth Five Year Plans resulted in a 60.3 per cent increase in known reserves. However, the per cent of the total survey drilling in the Soviet Union, allocated or conducted in this region, is small. [46] In the fourth Five Year Plan, 4 per cent of the survey drilling in the Union took place in Kazakhstan, and in the fifth, it decreased to 3.3 per cent.

The Turkmen Oil Regions.—Oil reserves in Turkmenistan comprise 2.17 per cent of the known reserves in the Soviet Union. The 200×10^6 tons of oil located in this republic have an energy potential of 2.3×10^{12} kilowatt hours.

Cheleken, south of Krasnovodsk on the Caspian coast, produced the first industrial oil in the Turkmen Region in the 1870's. [47] However, this region was relegated to a secondary position in the Republic when oil reserves in commerical quantities were discovered inland at Nebit-Dag in May of 1931. Nebit-Dag's first oil gusher yielded 500 tons of oil in 3 hours. [48] Previous discoveries, in the 1880's, were at shallow depths, approximately 130-160 meters, and were considered poor prospects.

During the fifth Five Year Plan, 1951-1955, over 61 per cent of the capital investment expended in drilling in these regions was for survey work. [49]

The Uzbek, Kirgiz, and Tadzhik A.S.S.R.'s.—Oil reserves in the

Eastern Republics of Middle Asia are centered in the Fergana Valley and the Tadzhik Depression. These adjacent regions contain an estimated 27×10^6 tons of oil. Uzbekistan is estimated to possess the major share of oil, 20×10^6 tons, with an energy potential of 232×10^9 kilowatt hours. This amounts to .22 per cent of the Union's total known oil reserves and their energy potential. Kirgizia, with approximately 5×10^6 tons of oil, has an energy potential of 58×10^9 kilowatt hours; known oil reserves, comprising .06 per cent of the Soviet total, are about equal to those of the Ukraine.

Tadzhikistan, with its Nefteabad oil fields, is estimated to have but 2×10^6 tons of oil, or .02 per cent of the known reserves in the Soviet Union. In his popularized geography book, Luknitski describes the oil well pumps as being tended by one woman in blue overalls. [50] His point was to depict the region's high degree of automation, but he has also conveyed an idea of the size of these operations and the extent of the field.

Ozocerite, also extracted by the Tadzhik oil trust, indicates the high paraffin base of this oil. [51] Deposits in the Tadzhik Depression have the highest sulfur content in oils of the Soviet Union. [52]

Survey drilling in the fifth Five Year Plan, in Middle Asia, increased 76 per cent over the volume of drilling in the fourth Five Year Plan. [53]

RÉSUMÉ ON THE ENERGY POTENTIAL AND QUALITY OF KNOWN RESERVES

Soviet Russia is possibly first among the oil endowed nations of the world. Current data indicate that in 1956 this vast subcontinent contained about 26 per cent of the world's proven oil reserves, not the 55 per cent set forth at the Geological Congress in 1937 and thereafter consistently claimed by Soviet writers. Free world crude reserves were reported as 26.7×10^9 metric tons on January 1, 1956. [54] If the Soviet reserves of 9.2×10^9 tons are added, the world total becomes 35.9×10^9 metric tons. Soviet Russia's share of these reserves is thus approximately 26 per cent. In 1956, the Middle East possessed about 53 per cent of the crude oil, North America 13 per cent, and South America an estimated 6 per cent. The remaining 2 per cent of the world's crude oil reserves was to be found in small deposits scattered throughout Europe, North Africa, and the Far East. However, the Soviet Union is undoubtedly entitled to claim first place in the reserves of crude oil among individual countries. Kuwait, the largest oil reserve region in the Middle East, possessed but 7.1×10^9 metric tons of oil in 1956, or 20 per cent of the world's proven crude oil reserves.

Nearly 96 per cent of the 106.9×10^{12} kilowatt hour energy potential of Soviet known oil reserves are located in European Russia, including the Caucasus and the west slope of the Urals. Unlike the distribution of coal, where 90 per cent of the reserves are located in remote regions east of the Urals, oil is found relatively near the major industrial complexes of the Union. This difference between the geographic distribution of the two energy resources will aid in the development of oil regions and consequently influence the industrial geography of the U.S.S.R.

With few exceptions, the quality of oil in European Russia is good. Quality is important to this study only when it directly affects the extraction of oil or is a contributing factor to the geographic location of a consuming enterprise. Information concerning the quality of Soviet oil and oil products is readily available, on even the most insignificant deposits. [55]

Some Ural-Volga oil deposits, notably those in Bashkiria, have a high sulfur content, higher than any other European oils, but generally not in excess of 3 per cent.

As previously mentioned, a high pour point of Ukhta oils creates extraction difficulties. Climatic conditions in this subarctic region, combined with the high pour point, make oil mining or shaft extraction necessary.

Ukrainian oils in the Borislav region must also be mined when the ozocerite content is found to be high. Mining in this instance is a secondary recovery measure. Imported oil products are more important in the economy of the Ukraine than locally produced oils. It is entirely probable that the value of the ozocerite alone makes this type of extraction feasible.

Contemplated mining of oil in the Baku Region is also a secondary recovery method, designed to procure the maximum amount of light oils from this region.

Emba oil in Kazakhstan, which comprises but 1 per cent of the total Soviet kilowatt hour energy potential, yields high quality aviation gasoline.

Sulfur contents in oils of the Tadzhik Depression range from 3 to 5 per cent. No other Soviet oils are known to have this high a sulfur content.

Because all Soviet crude oils have the same thermal value, 10,000 calories per kilogram, amount rather than quality is the important factor in determining the regional energy potential. According to the present status of surveying, the greatest potential is located in European Russia, principally the Ural-Volga Region. Caucasian deposits constitute a substantial second oil base. Both regions are so situated geographically that they can easily ship their oil to the established industrial regions of European Russia. Future developments may witness the Soviets supplying greater amounts of oil to European countries.

Before the Soviets can supply Europe with oil, however, they must meet the needs of the large South Siberian Belt if projected expansion in that region is to be completed. A virtual storehouse for hundreds of different minerals, this region lacks known oil reserves. Traces of oil exist in Siberia, but the quantity will remain unknown until proven by survey.

PROBLEMATIC OIL DEPOSITS

Numerous unsurveyed and undeveloped oil deposits exist throughout the Soviet Union. Many are recent discoveries, but others have been known for decades. Most of these deposits are located within the immense Siberian and Far Eastern territories, and all must be considered problematic until the extent of the reserves is proven by survey. To date, Soviet oil authorities have not published possible estimates on these deposits.

Geological surveys are in progress on the sedimentary basins of Siberia and the Far East. Available reports indicate that much of this work is of a reconnaissance nature, necessary before survey drilling is initiated. Keller reveals the actual amount of survey drilling which has taken place in this huge region when he states, [56]

> There has been a significant volume of survey drilling in regions of Siberia and the Far East. In the fifth Five Year Plan there were drilled 440.2 thousand meters, which is ten times more than in the fourth Five Year Plan (42.6 thousand meters.)

Percentages computed from these figures reveal that during the fourth Five Year Plan .06 per cent of the total survey drilling was carried out in these regions, and 3.8 per cent during the fifth Five Year Plan. Keller's use of the term *significant* describes the increase in survey drilling between five year plans, but under no circumstances should it be applied to the total volume of drilling in the region.

Exploratory work east of the Urals has taken place in Western Siberia. The combined search for oil and gas in the region began in 1930. [57] At the time, it was confined to the eastern part of the West Siberian Lowland. In 1932, I. M. Gubkin instituted formal study projects on oil prospecting in this region, the Kuznetsk Basin, and the Minusinsk Basin. From 1930 to 1948, most of the work was of a reconnaissance type, providing factual material which indicated prospective oil and gas deposits. Deep drilling was confined to the southwest part of the lowland and the northeast part of the Kuznetsk Basin. In 1949, survey drilling was co-ordinated with geophysical and hydrogeological surveying. Gas was discovered at Berezovo on the Ob

River, and some oil was obtained from the Jurasic beds in the Kolpashevo region of the Chelyabinsk Basin.

Two survey trusts operate in Western Siberia, the Zapsibneftegeologiya (West Siberian Oil Geology Trust) and the Sibneftegeofizika (Siberian Oil Geophysical Trust). The area to be covered is large, 3,400,000 square kilometers, and the strata deep, beyond 5 kilometers. [58] Western Siberia, preceded only by the Ural-Volga Region, ranks high in geophysical survey work. Despite this fact and the extended survey work accomplished in the past few years, there is little evidence that the geography of Soviet Russia's oil potential will be changed in the immediate future by present discoveries in this region.

Eastern Siberia has brighter prospects for the development of industrial oil supplies. Oil is known to exist in the Baikal Region, at Nordvik in the North, at Olenek and at the junction of the Tolba River with the Lena in Yakutsk, and in the Minusinsk Depression. [59] Survey work in Eastern Siberia is under the jurisdiction of the Vostsibneftegeologii (East Siberian Oil Geology Trust). Projected industrial developments should assure the intensification of oil prospecting, particularly in the southern part of the South Siberian Platform. [60] Other prospective sites in Siberia are the Ust Yenisei, the Minusinsk Depression, [61] and Southern Transbaikalia. [62]

European Russia also has unsurveyed and undeveloped oil deposits, in Belorussia and possibly in Moldavia. Belorussian deposits are an extension of the Dneper-Donets fields of the Ukraine. Oil in the Belorussian Republic is located at Elskaya. [63]

Major exploratory and survey efforts are concentrated in regions with a known energy potential. Problematic deposits, which could conceivably change this pattern, are still in the initial stages of exploration, and the energies expended for their development are slight. While some oil may be extracted from these deposits, it is on a local scale for local use. Soviet planners have outlined the development for Eastern Regions for the next decade, particularly for the South Siberian Belt. Oil is to play a part in this development, but it will be supplied from the Ural-Volga Region by pipeline. [64] Refineries under construction at Irkutsk will receive oil via an extension of the pipeline by way of Novosibirsk. [65]

The distributional pattern for oil production illustrates the importance of the various oil fields.

THE DISTRIBUTION OF OIL PRODUCTION AND CONSUMPTION IN THE U.S.S.R.

Crude oil production in the Soviet Union has increased by more than 1,000 per cent since 1913. Oil fields developed since the end of

World War II are now producing most of the crude oil. Deposits of the Ural-Volga Region produced over 70 per cent of the 113 × 10^6 tons of oil extracted in the U.S.S.R. in 1956. [66] In 1957, the Soviet Union produced 11 per cent of the world's crude oil, the United States 42 per cent, the Middle East 20 per cent, and Venezuela 14 per cent. [67] While Soviet production is well behind that of the United States, the use of oil in the Soviet economy is increasing.

Oil in the Structure of the Energy Economy.—Oil is less important in the energy economy of the Soviet Union than it is in the United States. In 1955, oil supplied but 9.2 per cent of the energy consumed in the Soviet Union (see Table 11). By comparison, oil in the United States supplied 43.6 per cent of the energy consumed in 1956. [68]

While use of oil in the Soviet Union may be low in comparison with the United States, it has, nevertheless, increased in the past two and a half decades. Per capita production of oil in 1913 was 66 kilograms; in 1957, it had risen to 483 kilograms. [69] As a producer of electricity, oil ranked slightly above peat in 1955, producing 8.98 per cent of the total electricity generated, or approximately 77 kilowatt hours per person out of a total per capita production of 850 kilowatt hours (see General Appendix Table I). The position of oil in today's energy economy in the Soviet Union is the result of an evolutionary process which began in the middle of the last century.

Trends in Production.—For nearly a century Soviet oil production has shown a steady increase. Baku and the North Caucasus accounted for most of the growth, until the rich Ural-Volga fields were exploited. Temporary reversals were occasioned by international and civil war at three different times. Production increased from 4 thousand tons in 1860 to 113 × 10^6 tons in 1958. [70] Growth in production was steady until 1901, when an output of 11 × 10^6 tons was attained. Between 1901 and 1905 production declined to 7 × 10^6 tons; internal political unrest and the Russian Japanese War were contributing factors. By 1916, production had again increased to nearly 10 × 10^6 tons; however, with the advent of the revolution and civil war, a drastic decline to 3.7 × 10^6 tons took place by 1921. From 1921 to World War II, production increased steadily; by 1941, the Soviets were producing 33 × 10^6 tons. [71] Warfare seriously interrupted production so that by 1943 the Soviets only produced 17.9 × 10^6 tons. The 1941 level of production was not gained again until 1949. The Azerbaidzhan-Baku region has never been able to achieve its prewar level of production.

Until World War II the vast preponderance of oil in the U.S.S.R. was produced in the Azerbaidzhan-Baku Region. In 1913, approximately 75 per cent of all oil came from this region; as late as 1940 it was still producing 71.5 per cent of the nation's oil. Major discoveries in the "Second Baku" during and after the war, and their subsequent development, resulted in the creation of a new oil center. Geographic patterns of oil production were entirely changed.

The influence of the Ural-Volga Region on the changing pattern of Soviet oil production is depicted in Table 15. A significant shift in production to this region did not take place until after the war. Since then it has surpassed the Baku Region, with the Bashkir Republic oc-

TABLE 15

OIL PRODUCTION IN THE URAL-VOLGA REGION, SELECTED YEARS 1940-1958, PROJECTED PRODUCTION 1960 (IN MILLIONS OF METRIC TONS AND PER CENT)[a]

Year	Total Soviet Production, Tons 10^6	Ural-Volga Production, Tons 10^6	Ural-Volga Production as a Per Cent of the Total Soviet Production
1940	31.1	1.8	6.0
1946	21.4	3.9	18.4
1950	37.9	11.0	29.0
1955	70.8	41.2	58.7
1956	83.6	52.8	63.0
1957[b]	100.0	68.0	68.0
1958[c]	113.0	80.0	70.0
1960[b] (Projected)	135.0	101.3	75.0

[a]A. A. Keller, *Neftyanaya i Gazovaya Promyshlennost S.S.S.R. V Poslevoennya Gody* (Moskva: Gostoptekhizdat, 1958), p. 15, for the years 1940-1956.

[b]A. A. Trofimuk, *Uralo-Povolzhe-Novaya Neftyanaya Baza S.S.S.R.* (Moskva: Gostoptekhizdat, 1957), p. 155.

[c]Estimated on the basis of data contained in: N. K. Baibakova, "Concerning the Plan for the Development of the National Economy of the R.S.F.S.R. in 1958," *Pravda*, No. 29 (14423), 29 January, 1958, p. 2, and Trofimuk, *op. cit.*, p. 155.

cupying first place in 1953, and the Tatar in 1957. [72] Production plans, projected to 1960, call for 75 per cent of the nation's total production from the Ural-Volga Region. If plans are completed in 1960, the Tatar Republic will produce 30 per cent of the nation's oil, followed by the Bashkir Republic and Kuibyshev Oblast, each of which will produce more than the Azerbaidzhan-Baku Region, which is scheduled for 11.7 per cent of the nation's production. [73] Technology, as well as discovery, aided the trend from Baku to the Ural-Volga Region.

Influence of New Technology on Soviet Production and the Development of New Regions.—Since it was first introduced in the Soviet Union, turbine drilling has become the accepted manner of drilling. Its use increased the total drilling from 6.5 per cent in 1946, to 87.6 per cent in 1956. [74] The turbine drills are able to penetrate the near crystalline hardness of the rocks, found especially in the Ural-Volga Region, at a much greater speed than conventional ones. [75] As a result, deeper geological strata can be reached in shorter time. The average operational well, in 1940, was 940 meters deep and the average survey well, 1,108 meters. [76] By 1956, operational drilling averaged 1,149 meters in depth and survey drilling averaged 1,996 meters.

By drilling deeper and faster, the turbine method advanced the development of the Ural-Volga Region more than any other technological discovery. Since the more productive strata in this region are the deeper strata, wells were projected downward through the Permian and Carboniferous to the Devonian, and more oil was obtainable. In 1940, when little turbine drilling was used and depths of operation were less than in 1956, the upper or Permian layers supplied 78 per cent of the Ural-Volga oil, and the Carboniferous 22 per cent. [77] By 1956, deeper drilling was prevalent in this region; at that time the Devonian supplied 73 per cent of the oil, the Carboniferous 22 per cent, and the Permian 5 per cent. Deep drilling was also instituted in the Kransodar and Baku Regions. Since the newly established technology and the newly discovered high oil potential regions have been incorporated into a producing economy, a stable regional pattern of oil production is now evident in the Soviet Union.

Regional Patterns of Production.—European Russia, including the Caucasus and west slope of the Urals, produced 90.64 per cent of the total 70.7×10^6 tons of oil in the Soviet Union in 1955 (see Fig. 14). Three regions in European Russia alone account for approximately 89 per cent of this production. The Ural-Volga Region produced 58.2 per cent of the Soviet oil, the Azerbaidzhan-Baku Region 21.6 per cent, and the North Caucasus 9.2 per cent (see Table 16). New deposits discovered in the past decade in the Krasnodar Krae account for much of the North Caucasus production. [78] Georgia contributed to the total production of the Caucasus Region by extracting 0.1 per cent of the Soviet oil. Collectively, the remaining two oil regions in the European part of the U.S.S.R. produce less than one and a half per cent of the Soviet output. Komi-Ukhta, in the North, extracts .78 per cent of the Russian oil, and the Ukraine .70 per cent.

In addition to the fact that European Russia contains 96 per cent of the known reserves, its geographical advantages also aid the Soviet economy. Oil is produced relatively near the major industrial centers. Established means of transportation exist so that the oil can be shipped easily. Because of present day overtaxed rail facilities, additional transportation facilities in the form of pipelines are under construction. Moreover, production and transportation are entirely within the Soviet boundaries and are therefore not subject to control over international waterway or to the problem arising from nationalism.

The oil output of the Kazakh-Emba Region in 1944 comprised 2 per cent of the total Soviet production. Remote from major consuming centers, this region ships out both crude and refined oil via pipeline and tanker. A pipeline carries oil to Orsk, and tankers from Guryev transport it, via the Caspian Sea, to the Volga River.

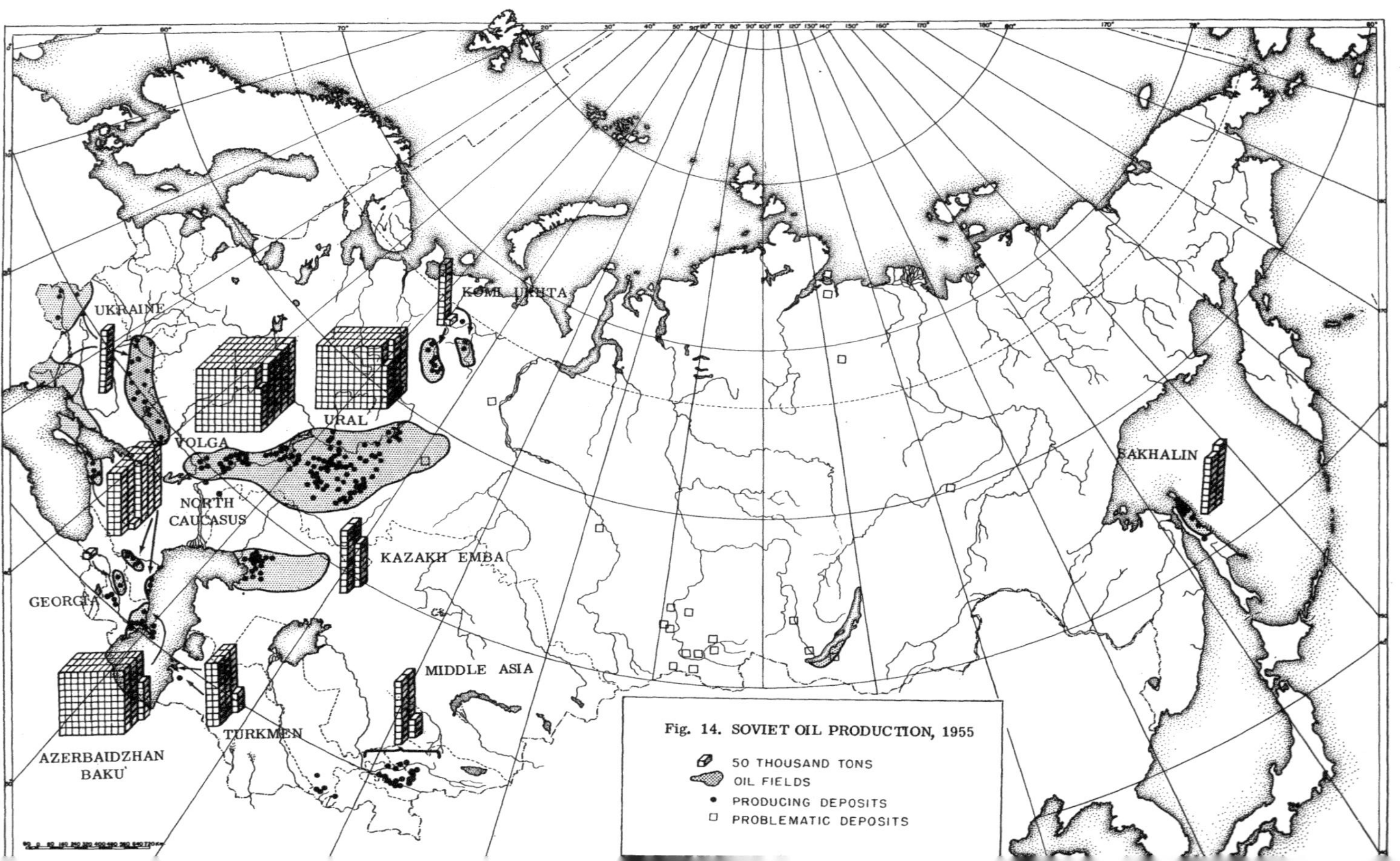

Fig. 14. SOVIET OIL PRODUCTION, 1955

TABLE 16

REGIONAL DISTRIBUTION OF SOVIET OIL PRODUCTION IN 1955, AND REPUBLIC PRODUCTION 1959 (THOUSANDS OF METRIC TONS AND PER CENT)

Region	Production Metric Tons 10^3	Per Cent of Total Production	Production Metric Tons 10^3 1959[h]
R.S.F.S.R. (incl.)	49,263.0	69.58	102,792
Total Ural-Volga[a]	41,218.0	58.22	---
Ural Sector[a] (incl.) Bashkir and Udmurt A.S.S.R.'s, Perm and Chkalov Oblasts	16,368.0	23.12	---
Volga Sector[a] (incl.) Tatar A.S.R., Ulyanovsk, Kuibyshev, Saratov, Stalingrad, and Penz[a] Oblasts	24,850.0	35.10	---
Komi-Ukhta[a]	553.0	0.78	---
Sakhalin[a]	950.0	1.34	---
North Caucasus[a]	6,542.0	9.24	---
Azerbaidzhan-Baku[b]	15,305.0	21.60	17,076
Georgia[c]	43.0	0.10	35
Ukraine[d]	531.2	0.70	1,627
Kazakh-Emba[e]	1,397.0	2.00	1,544
Middle Asia			
Turkmen[f]	3,126.0	4.40	4,577
Tadzhik[g]	16.8	0.02	17
Uzbek[b]	996.0	1.40	1,465
Kirgiz[b]	115.0	0.20	424
Total	70,793.0	100.0	129,557

[a]Computed from data in *World Oil*, Vol. 147, No. 3 (August, 13, 1958), 183.

[b]Ts. S.U., S.S.S.R. (K. G. Ivanova, ed.), *Promyshlennost S.S.S.R., Statisticheskiy Sbornik* (Moskva: Gosstatizdat, 1957), p. 155.

[c]Ts. S.U., S.S.S.R., Statisticheskoe Upravlenie Gruzinskoy S.S.R. (S. M. Pirumov, ed.), *Narodnoe Khozyaistvo Gruzinskoy S.S.R., Statisticheskiy Sbornik* (Tbilisi: Gruzglavpoligrafizdat, 1957), p. 42.

[d]Tsentralne Statistichne Upravlinnya pri Radi Ministriv S.R.S.R. Statistichne Upravlinnya Ukrainskoy R.S.R., *Narodnoe Gospodarstvo Ukrainskoe R.S.R., Statistechniy Zbirnyk* (Kiev: Derzhavne Statistichne Vidavnitsvo, 1957), p. 40.

[e]Statisticheskoe Upravlenie Kazakhskoy S.S.R., (A. Z. Grabarnik, ed.), *Narodnoe Khozyaistvo Kazakhskoy S.S.R., Statisticheskiy Sbornik* (Alma-Ata: Kazzgosizdat, 1957) p. 37.

[f]Ts. S.U., S.S.S.R., Statisticheskoe Upravlenie Turkmenskoy S.S.R., (A. Charyev, ed.), *Narodnoe Khozyaistvo Turkmenskoy S.S.R., Statisticheskiy Sbornik* (Ashkhabad: Gosstatizdat, 1957), p. 27.

[g]Tsentralnoe Statisticheskoe Upravlenie pri Sovete Ministrov S.S.S.R., Statisticheskoe Upravlenie Tadzhikskoy S.S.R., (S. E. Linderman, ed.), *Narodnoe Khozyaistvo Tadzhikskoy S.S.R., Statisticheskiy Sbornik* (Stalinabad: Gosstatizdat, 1957), p. 22.

[h]Tsentralnoe Statisticheskoe Upravlenie pri Sovete Ministrov S.S.S.R., (S. Ya. Genin, ed.), *Narodnoe Khozyaistvo S.S.S.R. V 1959 Gody, Statisticheskiy Ezhegodnik* (Moskva: Gosstatizdat, 1960), p. 186.

Cheleken and Nebit-Dag, along the east coast of the Caspian Sea in the Turkmen Republic, were responsible for 4.4 per cent of the Union's oil production in 1955. Oil is refined at Krasnovodsk and shipped by pipeline to Ashkabad. Geographically, this region has the distinction of being the largest oil producer east of the Caspian Sea.

The Fergana Valley and adjacent Tadzhik Depression produced 1.62 per cent of the nation's oil in 1955. Politically, this region is divided by a complexity of boundaries among the Uzbek. Kirgiz, and Tadzhik Republics. Foremost producer among these republics was the Uzbek with an output of 996,000 tons or 1.4 per cent of the Union's total. Only .2 per cent of the nation's oil was produced in the Kirgiz section of the Fergana Valley; oil in the Tadzhik fields comprised .02 per cent.

Industrial quantities of oil are not produced anywhere in the immense territory east from the Fergana Valley to the Pacific Ocean. Far Eastern oil production on an industrial basis is restricted to the Island of Sakhalin. In 1955, oil fields in the northern part of this island produced 1.34 per cent of the Russian oil.

Regional Patterns of Consumption.—Although specific data on the consumption of oil and oil products are not available, a reasonably accurate estimate of the regional distribution can be made by studying the operations of the Glavneftesbyta (Chief Trust for the Supply of Oil Products). The function of this trust is to supply Soviet consumers with oil and oil products. In fulfilling this duty, the trust maintains oil bases, or tank farms, in the consuming region. Data on the distribution of these tank farms provide an interpretative view of the regional distribution of consumers. The trust operates and maintains a total of 1,756 tank farms in the U.S.S.R. [79] European Russia, including the Urals and the Caucasus Regions, contains 1,243 of these tank farms. A total of 483 are located in the major oil producing areas: 111 are in the Urals, 168 in the Volga Region, 135 in the North Caucasus, and 69 in the Transcaucasus. Azerbaidzhan has only 29. Many of the tank farms in producing regions must be considered as storage depots at the point of production. Of the remaining 760 tank farms distributed throughout European Russia, 354 are located in the Central Region, 314 in the Ukraine and Moldavia, 67 in the Belorussia, 35 in the Baltic States, 49 in the Northwest, and 32 in the North.

Kazakhstan is served by 88 tank farms and Middle Asia by 96. Turkmenistan, one of the major oil producers of Middle Asia, has 13. Fifty-six tank farms are in the Uzbek Republic, 14 in the Kirgiz, and 13 in the Tadzhik. Brenner states that "A substanial disparity exists between the production and consumption of petroleum in the regions of Middle Asia and the Far East ... which makes necessary the bringing of large quantities of petroleum into these regions." [80] Therefore, it is safe to assume that these tank farms are maintained for local consumption and are not export storage depots.

Western Siberia possesses 132 tank farms and Eastern Siberia 69. The entire Far East contains only 37 tank farms.

Because the trust operating these tank farms must supply consumers, this distribution is a reflection of the geographical pattern consumption. Not only is European Russia the major producer of oil, it is also the major consumer; approximately 71 per cent of the tank farms are located in this region. The majority of the farms, not in producing regions, are distributed throughout the Central Region and the Ukraine; this distribution indicates that oil consumption there is very heavy.

Transportation of Oil.—Transportation facilities for oil and oil products are centered in the regions of production and consumption. European Russia has the greatest density of railroad, waterway, and pipeline networks in the U.S.S.R. In 1956, these three methods of transporting oil carried 148.2×10^9 ton kilometers of oil (see Table 17).

TABLE 17

TRANSPORTATION OF OIL IN THE SOVIET UNION, 1940, 1945, 1950, 1956, BY RAIL, PIPELINE, AND RIVERS (IN BILLIONS OF TON KILOMETERS AND PER CENT)[a]

	1940		1945		1950		1956	
Transportation	Tons	Per Cent	Tons	Per Cent	Tons	Per Cent	Tons	Per Cent
Railroads	36.0	69.4	24.0	72.7	52.0	75.5	112.0	75.6
Pipelines	3.8	7.3	2.7	8.2	4.9	7.1	20.5	13.8
River	12.1	23.2	6.3	19.1	12.0	17.4	15.7	10.6
Total	51.9	100.0	33.0	100.0	68.9	100.0	148.2	100.0

[a]Tsentralnoe Statisticheskoe Upravlenie Pri Sovete Ministrov S.S.S.R., (K. G. Ivanova, ed.), *Transport i Svyaz S.S.S.R. Statisticheskiy Sbornik* (Moskva: Gosstatizdat, 1957), adapted and computed from data on pp. 35, 119, 260.

Railroads transported 75.6 per cent of the ton kilometers of oil in 1956, pipelines 13.8 per cent, and waterways 10.6 per cent. Since 1940, the ton kilometers of oil transported by railroads have increased 183 per cent, the amount by pipeline 441 per cent, and the amount by waterways 30 per cent. As production increased, all transportation media were required to carry much more oil, as the above percentages indicate; however, the proportionate share transported by rail has remained approximately the same, while pipeline transportation of oil has increased at the expense of water transportation. Every effort is being made in the Soviet Union to increase the transportation of oil by pipeline.

Between 1943 and 1950, the length of pipeline in the U.S.S.R. increased from 4.4 thousand kilometers to 5.4 thousand. [81] By 1955, the Soviets had a total of 10.4 thousand kilometers of pipeline which was increased to 16.7 thousand kilometers at the end of 1959. Current plans call for a pipeline from the Urals to Irkutsk on Lake Baikal, [82] and from the Urals to Atbazar in Kazakhstan [83] (see Fig. 15). Pipelines are also being constructed from the Tatar Republic oil fields to Moscow. Regardless of the spectacular length of the single pipeline from the Urals to Eastern Siberia, the greatest density and length of pipeline is still in European Russia; this fact attests to the predominance in production and consumption in that region.

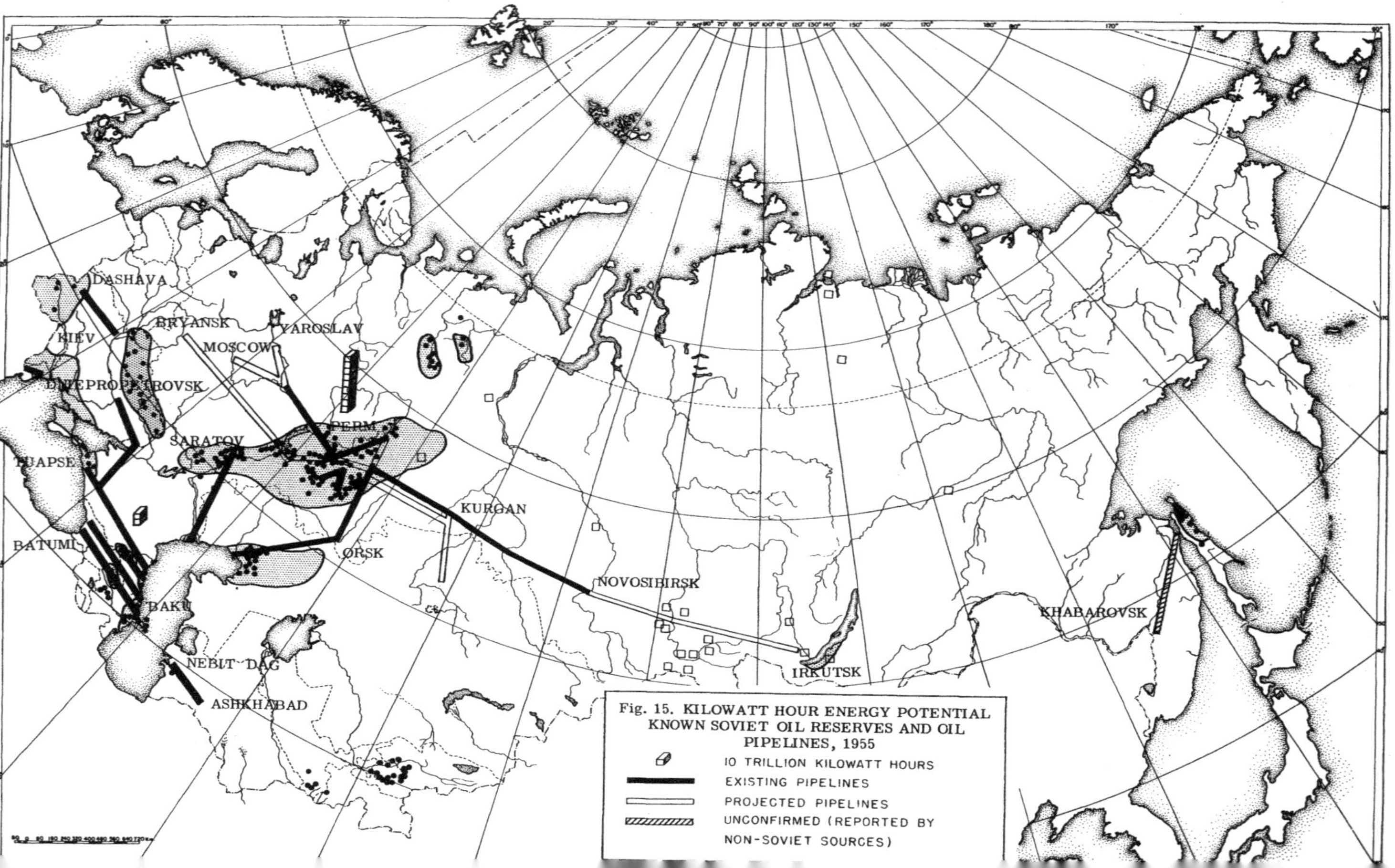

Fig. 15. KILOWATT HOUR ENERGY POTENTIAL KNOWN SOVIET OIL RESERVES AND OIL PIPELINES, 1955

NOTES – CHAPTER IV

1. A. Solodako, "Protiv Neddotsenki Nefti i Gaz v Ekonomike Strany," *Pravda*, No. 134 (13432), 14 May, 1955, p. 2.

2. G. D. Bakulev, *Voprosy Ekonomiki Topliva...*, *op. cit.*, p. 7.

3. Dr. J. Brian Eby, "How Oil Production in Russia Has Gained Since World War II," *World Oil* (Aug. 15, 1956), 182.

4. Computed from data in Eby, *op. cit.*, p..182, and Anon, "World Crude Reserves," *World Oil*, Vol. 145, No. 3 (Aug. 15, 1957), 198.

5. I. M. Goubkin, "World Petroleum Reserves," *Report of the XVII Session, International Geological Congress, Vol. I* (Moscow: 1939), 184.

6. A. G. Burenstam, "Otrazhenie Nekotorykh Voprosov Semiletnego Plana Razvitiya Narodnogo Khozyaistvo S.S.S.R., (1959-1965) v Kurse Geografii v Shkole," *Geografiya V. Skhole*, No. 2 (March-April, 1959), 5.

7. Academician A. V. Topchiev, "Razvivat Nauchnye Issledovaniya V Oblasti Organicheskogo Sinteza," *Pravda*, No. 157 (14551), 6 July, 1958, p. 2.

8. D. Notkin, "Perestroika Toplivnogo Balansa," *Planovoe Khozyaistvo*, No. 1 (1959), 44.

9. N. N. Kalinin, "40 Let Poiskov i Razvedki Neftyaykh i Gazovykii Mestrozhdeniy," *Razvedka i Okhrana Nedr*, No. 11 (1957), 42.

10. Anon. "Nekotorye Itogi i Perspekitivy Raboty Soundrakhozov Po Razvitiyu Neftyanoy Promyshlennost," *Nefti Khozyaistvo*, No. 2 (1958), 3.

11. For detailed discussions of reserve terminology see: J. V. Howell, *Glossary of Geology and Related Sciences* (Washington: The American Geological Institute, 1957), p. 243; Demitri B. Shimkin, *Minerals, A Key to Soviet Power* (Cambridge: Harvard University Press, 1953), pp. 19-22; Heinrich Hassman, *op. cit.*, p. 65; I. M. Goubkin, *op. cit.*, p. 184; and *Statistical Year Book World Power Conference, 1935*, p. 45.

12. Production computed from: Ts. S.U. S.S.S.R. (K. G. Ivanov, ed.), *Promyshlennost S.S.S.R.*, *op. cit.*, p. 154.

13. Anon, "World Crude Reserves," *World Oil*, Vol. 145, No. 3 (Aug. 15, 1957), 198. Barrels converted to metric tons.

14. Col. F. O. Miksche, "Geography and Strategy," in *The Red Army*, ed. by B. H. Liddell Hart (New York: Harcourt, Brace and Co., 1956), p. 242.

15. V. Kalamkarov, *op. cit.*, p. 18.

16. V. Kalamkarov, *op. cit.*, p. 19; and M. Brenner, "Problems of Petroleum in the Perspective of Development of the U.S.S.R. Economy," *Problems of Economics*, Vol. 1, No. 4 (August, 1958), 13. Translated contents of Feb. 1958, *Voprosy Ekonomiki* by International Arts and Sciences Press.

17. I. M. Goubkin (Gubkin), *op. cit.*, pp. 182-184; and S. F. Fedorov, *Petroleum Deposits of the Soviet Union* (Moskva: Gostoptekhizdat, 1939).

18. A. A. Keller, *Neftyanaya i Gazovaya Promyshlennost S.S.S.R. V Poslevoennya Gody* (Moskva: Gostoptekhizdat, 1958), p. 7.

19. A. A. Trofimuk, *Uralo-Povolzhe-Novaya Neftyanaya Baza S.S.S.R.* (Moskva: Gostoptekhizdat, 1957), p. 153.

20. V. D. Nalivkin, *et al.*, *Volga-Uralskaya Neftenosnaya Oblast' Tektonika* (Leningrad: Gostoptekhizdat, 1956), p. 27.

21. Z. L. Maimin, ed., *Ob Usloviyakh Obrazovaniya Nefti* (Leningrad: Gostoptekhizdat, 1955), p. 267.

22. Brenner, *op. cit.*, p. 13.

23. Ya. D. Gurevich, *et al.*, *Neftyanaya Promyshlennost S.S.S.R.* (Moskva: Gostoptekhizdat, 1958), p. 96.

24. A. A. Trofimuk, *op. cit.*, p. 198. This author does not agree with A. A. Keller, *op. cit.*, who lists 1,724.1 thousand meters for the 1946-1950 period.

25. S. N. Pavlova, et. al., *Nefti Vostochnykh Rainov S.S.S.R.* (Leningrad: Gostoptekhizdat, 1958), p. 248; and A. A. Trofimuk, *op. cit.*, p. 100.

26. Hassmann, *op. cit.*, p. 69.

27. I. P. Kuliev and Yu. A. Safarov, *Stroitelstvo Neftyanykii Skvazhin Na More* (Baku: Aznefteizdat, 1956), p. 9.

28. Gurevich, *et al.*, *op. cit.*, p. 117.

29. M. A. Kashkay and P. M. Alampiev, eds., *Azerbaidzhanskaya S.S.R. Ekonomiko-Geograficheskaya Kharakteristika* (Moskva: Geografgiz, 1957), p. 192.

30. Gurevich, *et al.*, *op. cit.*, p. 118.

31. L. Tairov, "Pokonchits Nedootsenkoy Sverkhglubokogo Bureniya," *Pravda*, No. 249 (13547), 6 September, 1955, p. 2.

32. M. A. Kashkay and P. M. Alampiev, eds., *op. cit.*, p. 194.

33. Kalamkarov, *op. cit.*, p. 19.

34. Keller, *op. cit.*, p. 46.

35. Keller, *op. cit.*, p. 12.

36. Keller, *op. cit.*, pp. 7-8.

37. A. G. Zadov, *et al.*, *The Oil Industry of Krasnodar* (Moskva: Gostoptekhizdat, 1957), p. 71. Abstracted in the Akademiya Nauk S.S.R., Institute Nauchnoy Informatsii, *Referativnyy Zhurnal, Geografiya* (Moskva: Proizvodstvenno-Izdatelskiy Kombinat VINITI, 1959), p. 271, Item No. 1963K.

38. T. T. Gonta, N. A. Gorev, I. F. Klitochenko, and K. F. Mikhailov, *Nefti Prirody Gaz Ukrainy* (Moskva: Gostoptekhizdat, 1957), p. 15.

39. Computed from data in Keller, *op. cit.*, p. 13 and Gonta *et al.*, p. 17.

40. Gonta, *op. cit.*, p. 64.

41. In 1937 Goubkin, *op. cit.*, p. 183, credited the region with only 10×10^6 tons; Federov, *op. cit.*, p. 530, in 1939, estimated but 1 million metric tons for the region. During the war years the region produced 1.2 million tons, Hassmann, *op. cit.*, p. 97. Devonian oil has been discovered since then.

42. A. F. Afanaseva, "30 Letie Neftyanoy Okhi," *Neftyanik*, No. 11 (1957), 23.

43. N. N. Pal'gov, *Kazakhstan* (Moskva: Gosudarstvennoe Izdatelstvo Geograficheskoi Literatury, 1953), p. 131.

44. S. V. Shumilin, "The Oil Fields of the Emba Region," *Abstract of Papers, International Geological Congress, XVII Session* (Moscow: 1937), p. 17.

45. I. G. Permiakov, "The Emba Salt Domes," *Abstract of Papers International Geological Congress, XVII Session* (Moscow: 1937), p. 17.

46. Keller, *op. cit.*, p. 8.

47. E. G. Freikin, *Turkmenskaya S.S.R.* (Moskva: Geografgiz, 1954), p. 224.

48. *Ibid.*, p. 228.

49. Computed from data in Keller, *op. cit.*, p. 46.

50. F. Luknitskiy, *Tadzhikstan* (Moskva: Molodaya Gvardiya, 1951), p. 144.

51. I. K. Narzikulova and S. N. Ryazantseva, eds., *Tadzhikskaya S.S.R., Ekonomiko-Geograficheskaya Kharakteristika* (Moskva: Geografgiz, 1956), p. 17.

52. Pavlova, *et al.*, *op. cit.*, p. 169.

53. Keller, *op. cit.*, p. 13.

54. Computed from data in *World Oil*, Vol. 147, No. 3 (August 15, 1958), 199.

55. An extensive pre-war survey of Soviet crude oils is found in A. S. Velekovskii, *Sovietskiye Nefty* (Moskva: Gos. Nauchnotekh, Izdatelstvo, 1938). Translated tables from this book are in Shimkin, *op. cit.*, pp. 206-207. The most recent compendium is Pavlova, *et al.*, *op. cit.* Over 450 pages of statistical data on Soviet oil and oil products are in this book. More than 92 oil deposits in the U.S.S.R. are covered, plus material on oil in the Mongolian Republic and China.

56. Keller, *op. cit.*, p. 13.

57. N. N. Rostovtseva, ed., *Geologicheskoe Stroenia i Perspektivy Neftegazonosnosti Zapadno-Sibirskoy Nizmennosti* (Moskva: Gosgeoltekhizdat, 1958), p. 3.

58. *Ibid.*, pp. 5 and 176.

59. V. G. Putsilo, M. N. Sokolova, and S. I. Mironov, *Nefti i Bitumy Sibiri* (Moskva: Izdatelstvo Akademii Nauk S.S.R., 1958) p. 8.

60. V. G. Vasilev, *et. al.*, *Geologicheskoe Stroenie Yuga Sibirskoi Platformy i Neftenosnost Kembriia* (Moskva: Gostoptekhizdat, 1957), 227 pp.

61. Vsesoyuznyy Nauchno-Issledovstelskiy Geologicheskiy Institut, Ministerstva Geologii Okharany Nedr S.S.S.R., *Geologichesky Stroenie S.S.S.R. Tom 1, Stratigrafiya* (Moskva: Gosgeoltekhizdat, 1958), pp. 304 and 474.

62. V. G. Putsilo, *et al.*, *op. cit.*, p. 31. The most recent work on Eastern Trans-Baikal is: V. N. Kozerenko, *Geologicheskoe Stroenie Yugo-Vostochnoi Chasti Vostochnogo Zabaikalya* (Lvov: Iv. Franko Gosundarstuennyy, Universitet, 1956), p. 308.

63. Pavlova, *et al.*, *op. cit.*, p. 198.

64. Trofimuk, *op. cit.*, p. 145.

65. S. Shchetinin, "Puti Razvitiya Proizvoditelnykh Sil Irkutskoy Oblasti," *Pravda*, No. 240 (14634), 28 August, 1958, p. 2.

66. *Pravda*, No. 330 (14724), 26 November, 1958, p. 1.

67. Computed from data in Anon. "World Crude Reserves," *World Oil*, Vol. 147, No. 3 (August, 1958), 135.

68. Computed from data in Kenneth E. Hill, Harold D. Hammar, and John G. Winger, *The Future Growth of the World Petroleum Industry* (New York: The Chase Manhattan Bank, 1957), p. 11.

69. L. Volodarsky, "Soviet Economy in a New Upsurge," *Problems of Economics*, Vol. 1, No. 11 (March 1939), 3. Translated contents of *Voprosy Ekonomiki*, Nov. 1958, by International Arts and Sciences Press.

70. Ts. S.U., S.S.S.R. (K. G. Ivanova, ed.), *Promyshlennost S.S.S.R.*, *op. cit.*, p. 153.

71. Keller, *op. cit.*, p. 14.

72. S. Ignatev, "Powerful Base for the Oil Industry," *Pravda*, No. 340 (13273), 29 December, 1954, p. 2; and Brenner, *op. cit.*, p. 12.

73. M. A. Evseenko, *Production of Oil in the U.S.S.R. in the Sixth Five Year Plan* (Moskva: Gostoptekhizdat, 1956), pp. 18-19.

74. Keller, *op. cit.*, p. 29.

75. Hassmann, *op. cit.*, p. 86.

76. Gurevich, *op. cit.*, p. 97.

77. Trofimuk, *op. cit.*, pp. 95 and 143.

78. Kalinin, *op. cit.*, p. 42.

79. Gurevich, *op. cit.*, p. 286.

80. Brenner, *op. cit.*, p. 13.

81. Tsentralnoe Statisticheskoe Upravlenie Pri Sovete Ministrov S.S.S.R. (K. G. Ivanova, ed.), *Transport V Svyaz SSSR Statisticheskiy Sbornik* (Moskva: Gosstatizdat, 1957), p. 260.

82. I. Lukin, "Chto Tromozit Stroitelstvo Transsibirskogo Neftepovoda," *Pravda*, No. 243 (14272), 31 August, 1957, p. 2.

83. *Promyshlenno – Ekonomicheskaya Gazeta*, No. 142 (1442), 30 November, 1958, p. 2.

Chapter V

NATURAL GAS: DISTRIBUTION OF RESERVES, ENERGY POTENTIAL, AND PRODUCTION

Geographical patterns for the distribution of natural gas reserves and their production began evolving in the late 1950's. Previously ignored, gas is presently undergoing intensive development, but is not yet an established major source of fuel in the U.S.S.R. If projected plans are successfully completed by 1965, gas will have achieved a position comparable in use as a supplier of energy to the gas industry of the United States.

Soviet claims, in 1958, listed possible gas reserves as 20×10^{12} cubic meters and known reserves as 10×10^{12} cubic meters. These figures should not be interpreted as absolute, because the infant Soviet gas industry is extremely dynamic. Even Russian authorities disagree in estimating possible reserves; a month prior to the publication of the previous figures, Vasilev and Elin listed foreseeable reserves as 18.3×10^{12} cubic meters. [1]

While the Soviets might possibly have reserves of 18-20 $\times 10^{12}$ cubic meters, this figure must be discarded as an unsound base upon which to compute energy potentials in favor of the known reserves of 10×10^{12} cubic meters. Possible reserves remain untested and unsurveyed, and their location is no more than speculation. More than 75 per cent of the known reserves were surveyed in the past five to six years, [2] but little more than 1×10^{12} cubic meters have actually been prepared for immediate industrial exploitation. [3] Known reserves of gas have increased approximately 1,000 per cent since the pre World War II period. [4] By way of contrast, the United States has 6.7×10^{12} cubic meters of surveyed industrial reserves, and the finding rate is 2.0 cubic feet per foot of gas withdrawn. [5]

Known gas reserves in the Soviet Union have an energy potential of 102.3×10^{12} kilowatt hours. This is less than either the energy potential for mineable reserves of oil shale or known reserves of oil.

The entire energy potential for known reserves of gas in the Soviet Union is approximately equal to the energy potential of the Ubagan brown coal basin in Kazakhstan (see Coal Appendix, Table II). However, it should be pointed out that, until recently, surveying for gas in the U.S.S.R. has been conducted at a slow rate. Future discoveries should remain well ahead of withdrawal rates, if promising regions prove as successful as known areas.

Between 1928 and 1955, gas production within the present boundaries of the Soviet Union increased 2,854 per cent. The phenomenonally high growth rate does not signify that gas occupies a position of prominence in the Soviet Union energy economy; rather it indicates the almost complete lack of gas in this energy economy in 1928.

SIGNIFICANCE OF NATURAL GAS IN THE SOVIET ECONOMY

Gas occupies a relatively insignificant position in the energy economy of the Soviet Union. On a thermal basis it supplies 2.3 per cent of the fuel balance of received energy resources and represents 6.5 per cent of the fuel balance of consumed energy resources (see Table 11). Projected plans for 1965 indicate that gas will supply approximately 25 per cent of the energy. [6] This would place it on a level comparable with the energy supplied by gas to the economy of the United States in 1955. It is not expected to exceed that level in the United States in 1965.

In the production of electricity, gas also occupied a relatively minor role in 1955. Gas accounted for 1.8 per cent of the electricity generated in that year, or 15.3 kilowatt hours per person. While this was twice the amount of electricity generated per person by shale, it was but one-fourth the amount generated by peat per person. (See General Appendix, Table I). In spite of these facts, thermoelectric stations are the largest consumers of gas in the Union. According to the 1957 plan, thermoelectric stations were to consume 30.6 per cent of the gas produced.

These thermoelectric stations actually consumed 34.6 per cent of the gas produced in 1958 (see Table 18). Projected plans for 1965, however, allocate but 20.5 per cent of the gas production to thermoelectric stations. While this is less of the total production, the quantity of gas consumed in thermoelectric stations will increase by three times as a result of the overall increase in gas production.

The domestic economy was the second largest consumer of gas in 1958. Gas consumed in this segment of the national economy will increase approximately three times in 1965. However, proportionately less of the total gas production will be used in this manner in 1965.

In 1958 the domestic economy consumed 13.2 per cent of the gas; in 1965 it will consume but 8.5 per cent.

The cement industry consumed 7.5 per cent of the gas in 1958. This industry is scheduled to consume five per cent of the gas in 1965—approximately four times the volume consumed in 1958.

Gas consumed as a raw material in the chemical industries accounted for 6.3 per cent of the production in 1958; it will consume but 5.8 per cent in 1965.

Metallurgical industries consumed 5.4 per cent of the gas in 1958. In 1965 they will consume 17.5 times as much gas as they did in 1958, or 18.8 per cent of the total production. No other segment of the economy is scheduled for such a large increase in the use of gas as is the metallurgical industry.

Soviet industries consumed 85.8 per cent of the gas produced in 1958, and if projected plans are successful they will consume 90.7

TABLE 18

CONSUMPTION OF GAS IN THE SOVIET UNION BY CONSUMING ENTERPRISE 1958, AND PROJECTED CONSUMPTION 1965[a]

	Volume of Gas Utilized in Millions[b] of Cubic Meters 10^6 and Per Cent				
Consuming Enterprises	1958 Millions Cu. M.	1958 %	Projected 1965 Millions Cu. M.	Increase 1959-1965 Number of times	1965 %
Total	28,846[c]	100.0	146,220[d]	5.1	100.0
Raw material for the chemical industry	1,880	6.3	8,500	4.5	5.8
Industrial consumers	26,685		136,520	5.1	
Including					
Metallurgical industry	1,568	5.4	27,500	17.5	18.8
Cement industry	2,160	7.5	7,500	3.5	5.1
Electro stations	9,971	34.6	30,000	3.0	20.5
Other industrial consumers	9,188	32.0	59,020	6.4	40.5
Domestic economy	3,798	13.2	12,500	3.3	8.5
Other consumers	281	1.0	1,200	14.4	.8

[a] Adapted and computed from D. I. Maslakov, *Toplivnyy Balans S.S.S.R.*, (Moskva: Gosplanizdat, 1960), p. 29.

[b] The original data in Maslakov was given as billions of cubic meters. This obviously was a typographic error.

[c] Other authorities do not agree with this total. A. F. Zasyadko in his book, *Toplivno—Energeticheskaya Promyshlennost S.S.S.R.*, (Moskva: 1959), p. 53, lists a total production of 28,084 million cubic meters for 1958, of which 22,512 million cubic meters were natural gas and 5,572 million cubic meters were incidental oil field gas. In the statistical handbook *Narodnoe Khozyaistvo S.S.S.R. V 1958 Gody*, natural and incidental gas production are listed as 28,084 million cubic meters and manufactured gas as 1,769.8 million cubic meters in 1958.

[d] A. F. Zasyadko, op. cit., p. 53 lists a projected figure of 148,280 million cubic meters for 1965.

per cent in 1965. Success in this endeavor cannot but materially aid Soviet industry, especially in lowering the cost of production and relieving the transportation system from shipping bulk fuels such as coal. Cost comparisons between natural gas, coal, and peat for Moscow indicate that this is true. Local peat shipped but 20 kilometers to Moscow is 1.5 times as expensive as natural gas shipped over 1,200 kilometers (see Table 19). Donets coal is 1.7 times as costly as natural gas when both are shipped to Moscow for consumption, and coal from the Moscow Basin transported from within 200 kilometers costs 2.1 times as much as natural gas shipped over 1,200 kilometers.

TABLE 19

THE COST OF NATURAL GAS COMPARED WITH COAL AND PEAT WHEN CONSUMED IN MOSCOW[a]

Type of fuel	Cost at the Place of Production Natural Gas equals 100	Distance Transported in Kilometers	Cost when Consumed in Moscow Natural Gas equals 100
Natural gas	100.0	1,200 – 1,300	100.0
Donets coal	730.1	1,100 – 1,200	172.5
Moscow coal	1,066.5	200	207.0
Local peat	644.7	20	153.4

[a]K. V. Dolgopolov, A. V. Sokolov, and E. F. Fedorova, *Neft i Gazy S.S.S.R.* (Moskva: Gosudarstvennoe Uchebno – Pedagologicheskoe Izdatelstvo Ministerstva Prosveshcheniya R.S.F.S.R., 1960), p. 98.

NATURAL GAS RESERVES IN THE SOVIET UNION

Gas in the known category, 10×10^{12} cubic meters, exists only in fields currently being exploited on an industrial basis. [7] These reserves are for pure gas and incidental oil field gas. Only Soviet data are available for estimating the possible gas content. According to Soviet sources, the gas content of their major oil fields averages approximately 10 per cent of the oil content. [8] The writer computed the 10 per cent content for each oil field, which gave the gas content in tons. This tonnage was then converted to cubic meters. A total gas content for all Soviet oil fields of 1.0×10^{12} cubic meters was estimated. It should be realized that the total is nothing more than an estimate made on an extremely generalized base. In amount, it is equal to the accumulated known reserves from two years' surveying work. Incidental gas reserves are included in the total figures for natural gas. This computation was made to determine their relative significance.

THE TOTAL ENERGY POTENTIAL OF NATURAL GAS

Known gas reserves in the Soviet Union contain a potential 102.3 $\times 10^{12}$ kilowatt hours, or 512 $\times 10^{3}$ kilowatt hours per person. Compared with the total 241,000 $\times 10^{3}$ per capita kilowatt hour energy potential of coal, the per capita energy potential of gas is insignificant. In total energy potential, it occupies fourth place among the non-renewable energy resources after coal, oil shale, and oil. Unlike the United States, where the energy content of natural gas is 140 per cent of the energy content of known oil reserves, [9] natural gas in the Soviet Union is but 96 per cent of the energy potential of known oil reserves.

Despite these facts, natural gas is destined for a prominent position in the Soviet energy economy because of its high thermal value per cubic meter, and the relative ease by which it can be transported. Sokolova claims that when extraction costs are measured for comparable units of energy, then gas is 12 times cheaper to produce than hard coal. [10]

The geographic location of gas fields in the Soviet Union further aids the importance of this energy resource.

DISTRIBUTION OF NATURAL GAS RESERVES AND THEIR KILOWATT HOUR ENERGY POTENTIAL

European Russia, excluding the Urals, possesses 87 per cent of the known reserves and 87 per cent of the kilowatt hour energy potential of natural gas (see Table 20). All of the remaining known reserves are located in the recently discovered Bukhara-Khiva gas fields of Uzbekistan. The regional distribution of natural gas reserves in European Russia can only be estimated, because Soviet sources do not yet contain an enumeration of this distribution.

Data indicate that the greatest concentration of known reserves of natural gas exists in the Ukraine, a direct result of intensive surveying, for both oil and gas, carried on in this region since 1950. Six Soviet geologists were recently awarded the "Lenin Prize" for the discovery and prospecting of "the largest gas field in the Union," the Shebelinka in the East Ukrainian gas region. [11] The inclusion of this field with other Ukrainian deposits brings the total kilowatt hour energy potential of the region to 25 per cent of the Union's total (see Fig. 16).

The Volga Region is the second largest gas province in the Soviet Union. Known reserves contain 20 per cent of the kilowatt hour energy potential. Industrially exploited deposits are located east of the Volga from Stalingrad north to Kuibyshev.

TABLE 20

ESTIMATED DISTRIBUTION OF KNOWN RESERVES OF NATURAL GAS AND THEIR KILOWATT HOUR ENERGY POTENTIAL

Region	Gas Reserves Cubic Meters[a] 10^{12}	Energy Potential Kilowatt Hours[b] 10^{12}	Per Cent of Energy Potential[c]
Ukrainian S.S.R.			
Western and Eastern fields[d]	2.5	25.5	25.0
North Caucasus[e]	1.5	15.4	15.5
Azerbaidzhan S.S.R.[f]	1.5	15.4	15.5
Volga-Ural[g]	2.0	20.4	20.0
Komi A.S.S.R.[h]	1.2	12.3	12.0
Bukhara-Khiva[i]	1.3	13.3	13.0
Total	10.0	102.3	100.0

[a]Data on the total known reserves from Yu. Bokserman, V. Kalamkarov, and A. Kortunov, "Zadachi Razvitiya Gozovoy Promyshlennost," *Planovoe Khozyaistvo*, No. 12 (1958), 30, were utilized in computing and estimating the distribution of known gas reserves.

[b]Computed in accordance with procedures outlined in introduction.

[c]Computed from data in Column 2.

[d]Estimated on the basis of data in: T. T. Gonta, N. A. Gorev, I. F. Klitochenko, and K. F. Mikhailov, *Neft i Prirodny Gaz Ukrainy* (Moskva: Gostoptekhizdat, 1957), p. 79; *Gazovaya Promyshlennost*, No. 3 (March 1956), 4-6; V. N. Kalchenko, "Prirodniy Gazer Palivnomu Balansi Ukrainskoi R.S.R.," *Visnik Akademii Nauk U.R.S.R.*, No. 9 (1957), 11-20; and production trends in Ts. S.U., S.S.S.R., (P. S. Bulgakov, ed.), *Narodne Gospodarstvo Ukrainskoi R.S.R. Statistichniy Zbirnik* (Kiev: Derzhstatvidav, 1957), p. 40.

[e]Estimated from data in: A. G. Zadov, *et. al.*, *Neftyanaya Promyshlennost Krasnodarskogo Kraya* (Moskva: Gostoptekhizdat, 1957), p. 71; *Gazovaya Promyshlennost* No. 12 (December 1956), 1-3; A. Karaev and G. Kovtunov, "Ryadom S Turboburom Kuvalda," *Pravda*, No. 23 (14624), 18 August, 1958, p. 2.

[f]Estimated from data in: G. Iskenderov, "Paths for Improving the Production and Utilization of Gas in Azerbaijan," *Planovoe Khozyaistvo*, No. 6 (1957), 80-83; M. A. Kashkay and P. M. Alampiev, *Azerbaidzhanskaga S.S.R.*, *Ekonomiko-Geograficheskaya Kharakteristika* (Moskva: Geografgiz, 1957), p. 195; S. S. Balzak, V. F. Vasyutin, and Ya. G. Feigin, *Economic Geography of the U.S.S.R.* (New York: The Macmillan Company, 1949), pp. 222-224; and production trends in Ts. S.U., U.S.S.R. (K. G. Ivanova, ed.), *Promyshlennost S.S.S.R.*, *Statisticheskiy Sbornik* (Moskva: Gosstatizdat, 1957), p. 156.

[g]Estimated from data in A. A. Trofimuk, *Uralo-Povolzhe-Novaya Neftyanoya Baza S.S.S.R.* (Moskva: Gostoptekhizdat, 1957), p. 183; *Neftyanoe Khozyaistvo*, Vol. 33, No. 4 (April 1955), 49-51 and production trends in Statisticheskoe Upravlenne R.S.F.S.R. (E. I. Froshenkov, ed.), *Narodnoe Khozyaistvo R.S.F.S.R.*, *Statisticheskiy Sbornik* (Moskva: Gosstatizdat, 1957), p. 22.

[h]Estimated from data in: A. Ya. Krems, *Perspektiny Neftyanoy i Gazovoi Promyshlennost Komi A.S.S.R.* (Syktyvkar: Komi Knizhnoe Izdatel-stvo, 1957), p. 32; N. I. Shiskin, *Komi A.S.S.R.*, *Ekonomiko-Geograficheskaya Kharakteristika* (Moskva: Geografgiz, 1959), p. 31-33; and production trends in Statisticheskoe Upravlenie Komi A.S.S.R. (I. I. Tarabukin, ed.), *Narodnoe Khozyaistvo, Komi A.S.S.R. Statisticheskiy Sbornik* (Syktyvkar: Komi Knizhnoe Izdatelstvo, 1957), p. 22.

[i]G. Kh. Dikenshtein, *et. al.*, *Gazlinskoe Gazoneftyanoe Mestorzhdenie* (Moskva: Gostoptekhizdat, 1959), p. 45; S. Ziyadullaev, "Razvitie Narodnogo Khozyaistvo Uzbekskogo S.S.R. v 1959 – 1965 Godokh," *Planovoe Khozyaistvo*, No. 1 (1959) p. 66; *Pravda*, No. 270 (14664), 27 September, 1958, p. 2.

Gas fields of the Northern Caucasus are not depicted on Soviet maps as a continuum of the Volga gas province. [12] Approximately 15 per cent of the kilowatt hour energy potential of Soviet gas reserves are found in this region. Major deposits are located at Krasnodar and Stavropol.

The gas fields of Azerbaidzhan also contain about 15 per cent of the Soviet gas reserves. Karadag, the major deposit, has test wells with a reported depth of 3,823 meters, which are said to have produced gas under a pressure of 400 atmospheres. [13]

Natural gas reserves in the Komi A.S.S.R. are estimated at 1.2×10^{12} cubic meters with an energy potential of 12×10^{12} kilowatt hours. These figures, however, are a provisional estimate subject to revision when surveying is completed. Although the region may now be credited with 12 per cent of the Soviet kilowatt hour energy potential of natural gas, future prospects for substantial increases appear extremely favorable. A. Ya. Krems lists several prospective large gas bearing regions which have not yet been surveyed. [14] Foreign students have for some time considered the gas reserves in the Komi Republic more important economically than its oil reserves. [15]

Known reserves of natural gas in the Bukhara-Khiva field are reported by 1.3×10^{12} cubic meters. [16] The kilowatt hour energy potential comprises 13 per cent of the Soviet total. Surveying in this region, however, is far from completion. At the present time, the Soviets have directed their energies in preparing this field for exploitation, toward one major deposit, the Gazli. This deposit is reported to have 438×10^9 cubic meters of natural gas distributed throughout six gas bearing horizons. [17] Oil in small quantities is also obtained from the Gazli. Located on the edge of the Kyzyl-Kum Desert, it is situated 140 kilometers northwest of Bukhara.

In addition to the 10×10^{12} cubic meters of surveyed reserves of natural gas, the Soviets claim to have 10×10^{12} cubic meters of possible reserves in two major gas fields. Western Siberia contains one which is located along the Ob River near its mouth, [18] and the other is in the Yakutsk A.S.S.R., north and east of the city of Yakutsk. A third Siberian gas field is reported to exist in the Buryat A.S.S.R., but its location remains ill-defined. [19] While these deposits deserve mention, they are at present too obscure to include in a study of this nature.

DISTRIBUTION OF NATURAL GAS PRODUCTION

Trends in Production.—Gas production in the Soviet Union increased 2,854 per cent between 1928 and 1955 (see Table 21). This growth signifies not the intensive use of gas today but the almost complete lack of it in the Soviet energy economy of the earlier years.

Geographically, gas production has always been centered in the European section of the U.S.S.R., primarily the North Caucasus and

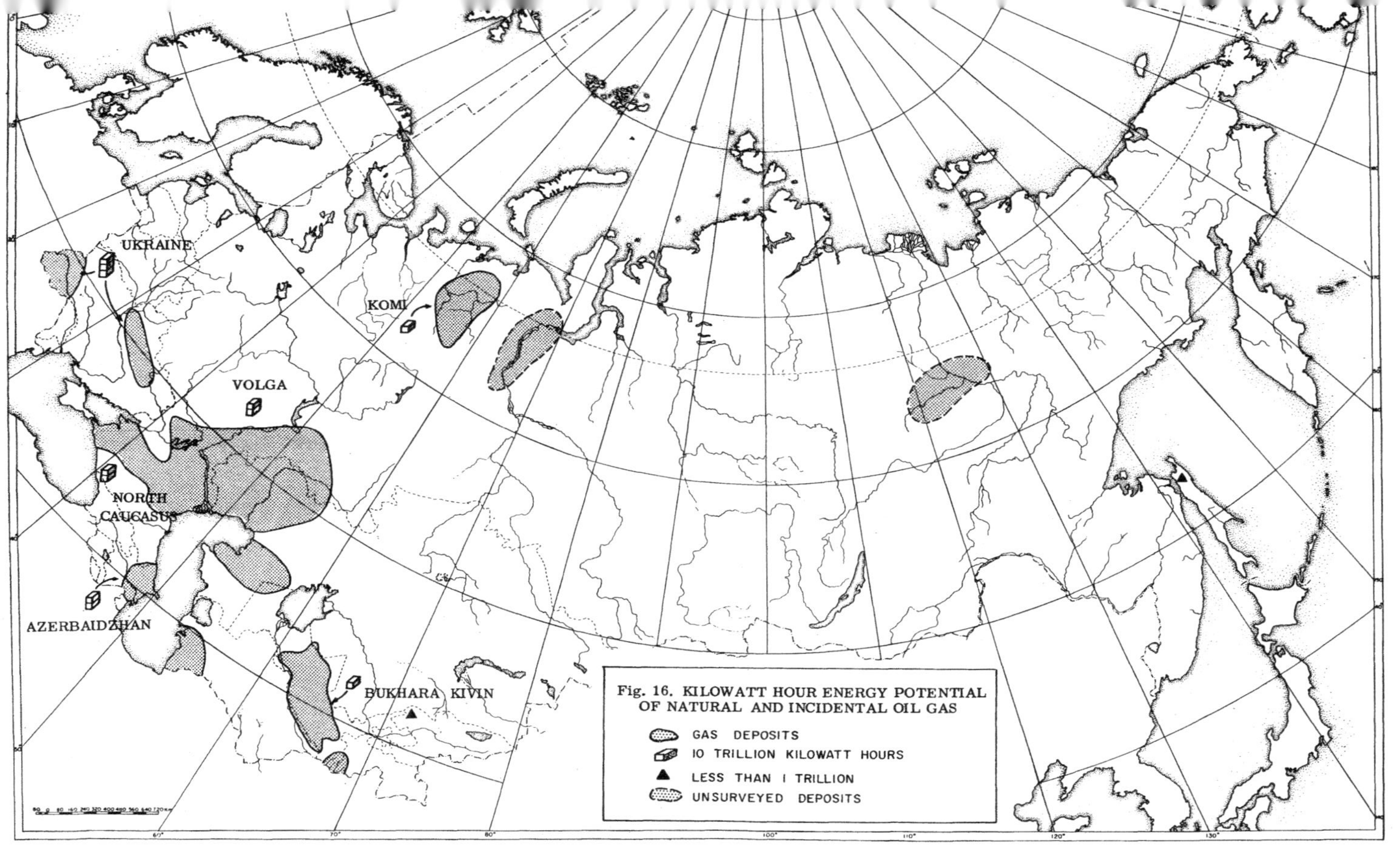

Fig. 16. KILOWATT HOUR ENERGY POTENTIAL OF NATURAL AND INCIDENTAL OIL GAS

Azerbaidzhan-Baku Regions. Until 1950, the Azerbaidzhan Region was the dominant producer. During the war, discovery and exploitation of gas deposits in the Saratov Region facilitated large scale production after the cessation of hostilities. Post war development witnessed an upsurge in gas production in the Komi A.S.S.R. and the establishment

TABLE 21

PRODUCTION OF NATURAL GAS, SELECTED YEARS 1928-1955 (IN MILLIONS OF CUBIC METERS)[a]

Region	1928	1932	1937	1940	1945	1950	1955	1958[b]
Total U.S.S.R.	304.0	1,049.0	2,178.9	3,219.1	3,378.0	5,760.9	8,980.0	28,084.5
R.S.F.S.R.	126.9	519.8	180.0	209.9	1,494.8	2,867.3	4,291.0	13,743.4
North	---	---	---	---	447.0	1,076.4	1,075.9	1,139.0[c]
Volga Region	---	---	---	---	728.4	964.7	1,627.2	5,164.0[c]
North Caucasus	126.9	519.8	156.3	195.0	120.2	322.2	595.3	5,851.0[c]
Urals	---	---	23.7	14.9	130.9	418.9	799.8	1,318.0[c]
Far East	---	---	---	---	68.4	85.1	192.8	271.0[c]
Ukrainian S.S.R.	---	---	---	495.1	776.9	1,536.5	2,927.6	9,500.7
Uzbek S.S.R.	---	3.0	1.0	0.7	8.9	52.2	103.0	126.4
Kazakh S.S.R.	1.3	2.3	3.7	3.9	4.9	7.4	24.7	42.0
Azerbaidzhan S.S.R.	175.5	522.2	9,991.0	2,498.1	976.8	1,232.8	1,493.8	4,446.2
Kirgiz S.S.R.	---	---	---	---	0.1	---	---	1.5
Tadzhik S.S.R.	0.3	1.9	3.2	2.2	0.8	0.2	---	---
Turkmen S.S.R.	---	---	---	9.2	14.9	64.5	140.8	224.3

[a]Ts. S.U., S.S.S.R. (K. G. Ivanova, ed.), *Promyshlennost S.S.S.R., Statisticheskiy Sbornik* (Moskva: Gosstatizdat, 1957), p. 156.

[b]Tsentralnoe Statisticheskoe Upravlenie Pri Sovete Ministrov S.S.S.R., *Narodnoe Khozyaistvo S.S.S.R. V 1958 Gody Statisticheskiy Ezhegodnik* (Moskva: Gosstatizdat, 1959), p. 213.

[c]Tsentralnoe Statisticheskoe Upravlenie Pri Sovete Ministrov R.S.F.S.R., *Narodnoe Khozyaistvo R.S.F.S.R. V 1958 Gody* (Moskva: Gosstatizdat, 1959), p. 90.

of Ukrainian fields as the major producer. During this entire period, most of the gas produced was consumed locally. One major pipeline from Saratov to Moscow existed for the supply of distant centers.

Between 1955 and 1958, gas production increased from 8.9×10^9 cubic meters to 30×10^9 cubic meters, or 237 per cent. [20] Scheduled production by 1965 is 150×10^9 cubic meters. [21] The 1965 gas production will be equal in thermal value to the combined coal production of the Donets, Moscow, and Pechora Coal Basins. [22] If this production is achieved, then gas, as a supplier of energy in the U.S.S.R., will have attained the same status that it has in the fuel supply of the United States.

The Regional Pattern of Gas Production in 1955.—European Russia, including the West slope of the Urals and the Caucasus, produced 8.5×10^9 cubic meters of gas in 1955, or approximately 95 per cent of the total Soviet output (see Fig. 17). Practically all of this gas was utilized locally because interregional pipelines were few in number.

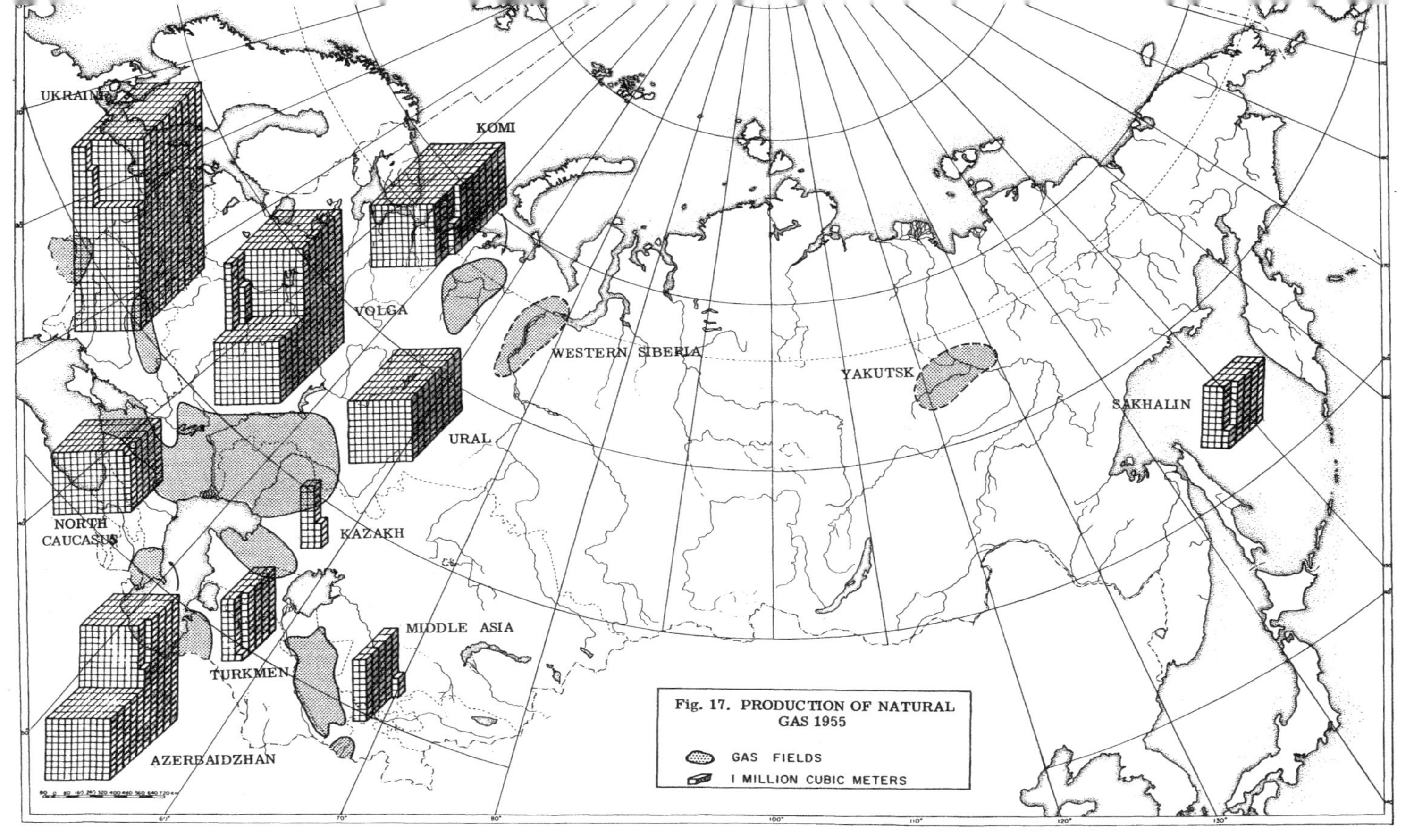

Fig. 17. PRODUCTION OF NATURAL GAS 1955

Foremost among the European producers was the Ukraine, in 1955 producing nearly 33 per cent, or 2.9×10^9 cubic meters of the natural gas. However, by 1956, the Ukraine produced 4×10^9 cubic meters of gas. [23] When all consumed fuels are measured in terms of their thermal value, gas accounted for 3.6 per cent of the energy consumed in the Ukraine in 1956. [24] Planned production for 1960 hopes for a gas output of 13.3×10^9 cubic meters for the Ukraine. [25] Thus in the five year period, 1955 to 1960, gas production was scheduled for a 393 per cent growth. In 1965, the Shebelinsk deposit alone is expected to produce 10×10^9 cubic meters of gas, which is equal in calorie content to 13×10^6 tons of Donets coal. [26]

Gas production in the combined Volga-Ural Region accounted for approximately 27 per cent of Soviet production in 1955. Production in the Volga section was 1.6×10^9 cubic meters or 18 per cent of the entire Union's output. In the Urals section, the Soviets produced 799×10^6 cubic meters of gas, or nearly 9 per cent of the total gas production in the country.

Azerbaidzhan ranked third in gas production in 1955, with 16.6 per cent of the total Soviet output. Production was scheduled to increase from 1.4×10^9 cubic meters in 1955 to 6.0×10^9 in 1960. [27] The Karadag deposit, which produced 27.6 per cent of the gas in this region in 1956, is expected to supply the major portion of the increase.

Only 6.6 per cent of the total Soviet output of natural gas in 1955 occurred in the North Caucasus Region. Here reserves equal, and possibly exceed, those of Azerbaidzhan; yet production is much less. It is possible that full scale exploitation of these reserves is being deferred until the interregional pipelines to Moscow and Leningrad are completed. Even so, Zadov states that gas production in 1956 in the Krasnodar Krae alone, was equivalent in energy content to 800×10^3 tons of hard coal. [28]

The remaining 5 per cent of the Soviet gas production in 1955 occurred east of the Urals in four separate deposits. Incidental gas from oil deposits accounted for most of this production. Deposits on Sakhalin Island were responsible for 2.1 per cent of the Union's output; gas in this region is used by the oil industry and in the city of Oka where it accounts for 80 per cent of the fuel. [29] Approximately 1.6 per cent of the Soviet total was attributed to incidental gas production from oil deposits in the Turkmen Republic; this gas is utilized in workers settlements and the city of Nebit Dag. [30] The large gas deposits of Southwestern Turkmenistan have not been exploited on an industrial basis. In 1955, gas production in the Kazakh Republic was but .3 per cent of the total output in the U.S.S.R.; Emba oil deposits were responsible for this production. The gas field which lies between the Caspian Sea and the Aral Sea remains unsurveyed. Uzbekistan produced the remaining 1.1 per cent of the natural gas in

1955. This production was not from the recently discovered Bukhara-Khiva fields, but incidental oil field output.

The Emerging Importance of Natural Gas in the U.S.S.R.—Natural gas is just now emerging as an important source of energy in the Soviet Union. Future production plans show a prominent place for it in the energy balance of consumed fuels in the nation. By 1965, natural gas is expected to provide approximately 25 per cent of the total energy (see Table 11). In the Urals Region alone, gas will increase from 0.9 per cent to 28.1 per cent in the balance of consumed energy resources (see Table 12).

The Soviets' reasons for emphasizing natural gas in their energy economy are clear, and in discussing the advantages of this fuel, M. A. Kashkay and P. M. Alampiev state,

> The production of natural gas is economically very effective since the productivity of labor in the output of gas is 31 times greater than in the production of peat, 21 times greater than in the production of oil shale, 12 times greater than in the production of coal, and 2-3 times higher than in the production of oil. [31]

In order to attain these advantages, gas must be brought to the Soviet consumer. The present Seven Year Plan calls for the gasification of 160 cities, [32] but first the planned pipeline network now under construction must be completed. This network of 26 thousand kilometers is scheduled for installation between 1959 and 1965 [33] (see Fig. 18).

Regions benefiting from this pipeline network are the Central Region around Moscow, the Leningrad Region, the Ukraine, the Caucasus, the Urals, and Central Asia. Gas is to be supplied to the Urals from the large Bukhara-Khiva deposit in Uzbekistan, the Volga deposits, and the deposit at the mouth of the Ob River in Western Siberia. Gas will be shipped from the North Caucasus deposits to Moscow and Leningrad and from the Ukraine to these centers.

Correlative Summary on Reserves and Production.—The preponderance of Soviet gas reserves and gas production is in European Russia. Future developments in exploratory surveying may, and probably will, result in a shift in the reserve base. It is entirely feasible to expect the vast sedimentary basins east of the Urals to yield gas. However, the reserves now located and known in European Russia are sufficient to satisfy production demands for this type of fuel for several decades.

At the present stage in the development of the Soviet gas industry, the patterns for the distribution of gas reserves must be considered as tentative. Production patterns will also undergo change as new reserve bases are developed. These developments are in the immediate future, because several years are required to survey and install equipment necessary for the exploitation of an oil field.

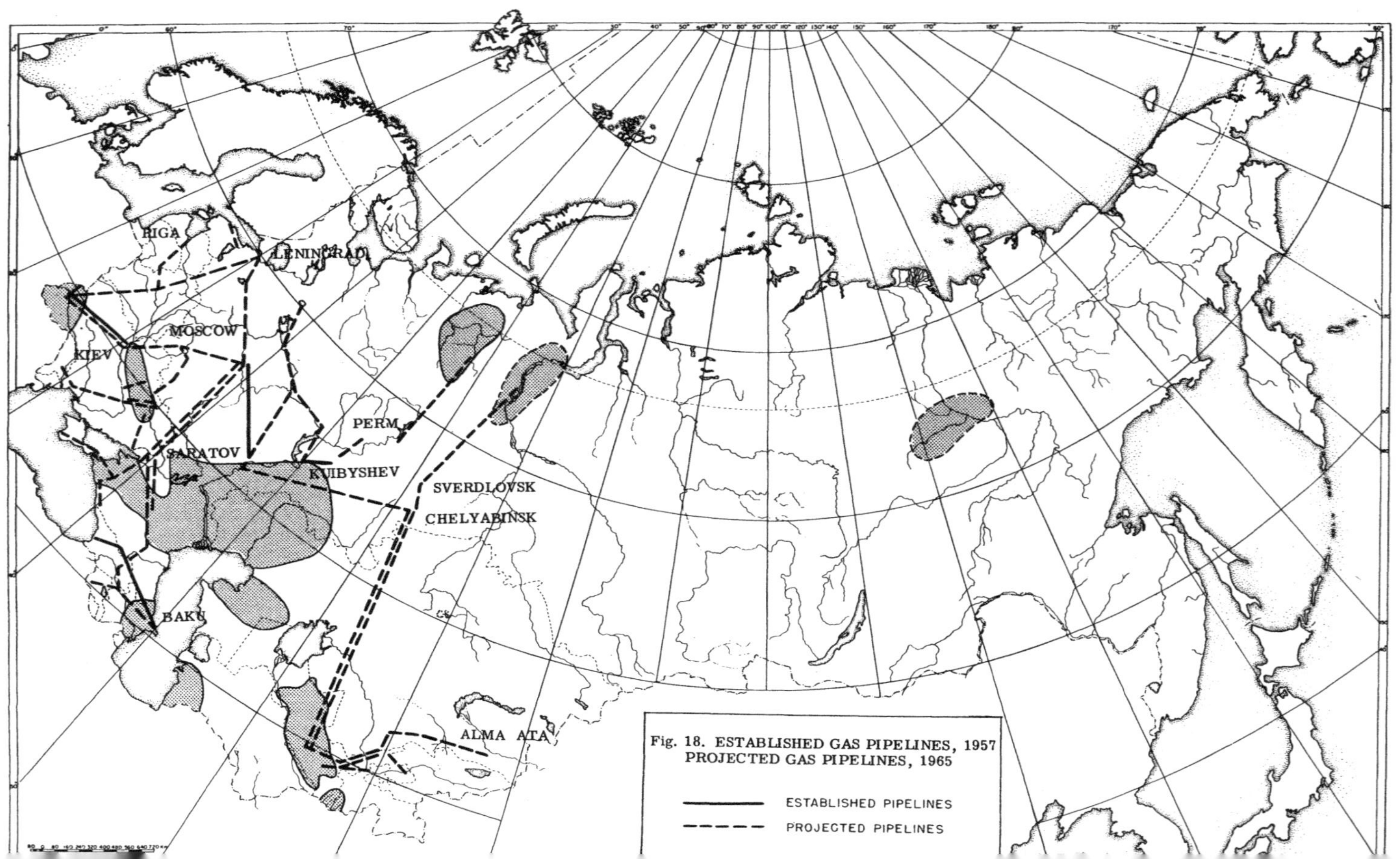

Fig. 18. ESTABLISHED GAS PIPELINES, 1957
PROJECTED GAS PIPELINES, 1965

NOTES – CHAPTER V

1. Yu. Bokserman, V. Kalamkarov, and A. Kortunov, "Zadachi Razvitiya Gazovoy Promyshlennost," *Planovoe Khozyaistvo*, No. 12 (1958), 30; V. G. Vasilev and N. D. Elin, "Prognoznye Zapasy Gaza Sovetskogo Soyuza," *Gazovaya Promyshlennost*, No. 11 (1958), 1-4. See also *Izvestiya Akademiya Nauk S.S.S.R.*, *Seriya Geog.*, July-August, 1959, pp. 46-54.

2. N. S. Erofeev, "Razvedka Mestorozheniy Prirodnogo Gaza i Ee Perspektivy," *Gazovaya Promyshlennost'*, No. 11 (1957), 9.

3. Bokserman, *et al.*, *op. cit.*, p. 30.

4. Computed from data in Balzak, *et al.*, *op. cit.*, p. 222, and Bokserman, *et al.*, *op. cit.*, p. 30.

5. Lyon F. Terry and John G. Winger, *Future Growth of the Natural Gas Industry* (New York: The Chase Manhattan Bank, 1957), p. 15.

6. E. Sokolova, "The Structure of the Fuel Balance," *Problems of Economics*, Vol. 1, No. 7 (1958), 33. Translated contents of *Voprosy Ekonomiki*, May 1958, by International Arts and Sciences Press.

7. Bokserman, *et al.*, *op. cit.*, p. 30.

8. Balzak, *et al.*, *op. cit.*, p. 222.

9. Terry and Winger, *op. cit.*, p. 17.

10. Sokolova, *op. cit.*, p. 33.

11. *U.S.S.R.*, No. 7 (34), 1959, p. 45.

12. *Promyshlenno-Ekonomicheskaya Gazeta*, No. 142 (442), 30 November, 1958, p. 2.

13. G. Iskenderov, *Planovoe Khozyaistvo*, No. 6 (1957), 80.

14. A. Ya. Krems, *Perspektivy Neftyanoy i Gazovoi Promyshlennost Komi A.S.S.R.* (Syktyvkar: Komi Knizhnoe Izdatelstvo, 1957), pp. 22-23.

15. Hassmann, *op. cit.*, p. 97.

16. S. Ziyadullaev, "Razvitie Narodnogo Khozyaistva Uzbekskogo S.S.R. v 1959-1965 Godakh," *Planovoe Khozyaistvo*, No. 1 (1959), 66.

17. G. Kh. Dikenshtein, *et al.*, *Gazlinskoe Gazoneftyanoe Mestorzhdenie* (Moskva: Gostoptekhizdat, 1959), p. 45. *The Kiev rabochaya Gazeta*, for October 1959 gives the Gazli deposit as 460×10^9 cubic meters.

18. Rostovtseva, ed., *op. cit.*, p. 251.

19. *Geologiya Nefti*, No. 1 (1957), 2. *Pravda*, on Oct. 27, 1959 reported a successful gas well at Osinsky in Irkutsk Oblast.

20. Bokserman, *et. al.*, *op. cit.*, p. 28.

21. *Pravda*, No. 242 (14636), 30 August, 1958, p. 1.

22. *Pravda*, No. 189 (14583), 8 July, 1958, p. 1.

23. Gonta, *et al.*, *op. cit.*, p. 75.

24. V. N. Kalchenko, "Prirodniy Gazo Palivnomu Balansi Ukrainskoi R.S.R.," *Visnik, Akademii Nauk U.R.S.R.*, No. 9 (1957), 12.

25. *Ibid.*, p. 12.

26. V. Teslenko, "Shebelinskiy Gaz-Narodnony Khozyaistuv," *Pravda*, No. 251 (14655), 18 September, 1958, p. 1.

27. Iskenderov, *op. cit.*, p. 80.

28. Zadov, *et al.*, *op. cit.*, p. 4.

29. Afanaseva, *op. cit.*, p. 26.

30. Freikin, *op. cit.*, p. 185.

31. Kashkay and Alampiev, *op. cit.*, p. 195.

32. *Promyshlenno – Ekonomicheskaya Gazeta*, No. 142 (442), 30 November, 1958, p. 2.

33. *Promyshlenno – Ekonomicheskaya Gazeta*, 31 August, 1958, p. 1.

Chapter VI

THE TOTAL ENERGY POTENTIAL OF ALL NONRENEWABLE ENERGY RESOURCES IN THE SOVIET UNION

When all energy resources are viewed in retrospect, the potential energy content of coal makes it the most important nonrenewable energy resource in the Soviet Union and in the world. Computations based on Soviet reserve data reveal that coal alone accounts for approximately 99.3 per cent of the total $46{,}562.8 \times 10^{12}$ kilowatt hour energy potential of all nonrenewable resources in the U.S.S.R. (see Table 22). While this figure may seem exceedingly large, it is only 1.9 per cent more than the ratio of coal to all other nonrenewable resources, when similar comparisons are made for the world. Comparisons with data on nonrenewable resources of the United States reveal that coal accounts for approximately 95 per cent of the total energy potential. Thus, there is but a four per cent difference between the position of coal in the energy balance of the nonrenewable resources in the United States and the Soviet Union.

Comparative data on the position of oil shale indicate that its ultimate energy potential is far less significant in the Soviet Union than it is in the United States. Its development as a fuel in the Soviet Union is well advanced.

The energy potential of oil in the Soviet Union is 0.23 per cent of the total kilowatt hour energy potential of all nonrenewable resources, while in the United States it comprises 1.0 per cent. In both countries and throughout the world, oil has a lower energy potential than oil shale.

Natural gas possesses 0.22 per cent of the kilowatt energy potential of all nonrenewable resources in the Soviet Union. In the United States it accounts for 0.69 per cent of the total energy potential.

The dominant position of coal in the energy potential of nonrenewable reserves should not negate the importance of the practical day to day exploitation of all other fuels. Moreover, the nature of this

Table 22

A COMPARISON OF THE TOTAL ENERGY POTENTIAL OF ALL CONVENTIONAL NONRENEWABLE ENERGY RESOURCES OF THE U.S.S.R., THE UNITED STATES, AND THE WORLD

Resources	U.S.S.R. Energy Potential Kwt. Hrs. 10^{12} [a]	U.S.S.R. Energy Potential Per Cent[b]	United States Energy Potential Per Cent[c]	World Energy Potential Per Cent[d]
Coal	48,233.1	99.29	94.66	97.40
Oil Shale	120.1	0.26	3.65	1.56
Oil	106.9	0.23	1.00	0.65
Natural Gas	102.3	0.22	0.69	0.39
Total	48,562.4	100.00	100.00	100.00

[a] Assembled from data in Chapters I, III, IV, and V.

[b] Computed from data in Column 1.

[c] Eugene Ayres and Charles A. Scarlott, *Energy Sources – The Wealth of the World* (New York: McGraw-Hill Book Company, Inc., 1952), p. 84. Percentages were computed from absolute data presented in horsepower hours, maximum values.

[d] *Ibid.*, p. 83.

study and its material result in the use of superlatives: billions of tons of oil, trillions of tons of coal, quadrillions of kilowatt hours. These figures may well create unwarranted confidence in the Soviet energy resource base, because such sums in no way imply recoverability. Experience throughout the world does indicate, however, that a 50 per cent recovery may be the maximum for coal and 20 per cent for oil. Despite their ultimate recoverability, the regional distribution of these energy resources currently remains the important factor.

THE REGIONAL MAP OF SOVIET ENERGY POTENTIALS

The distributional pattern of kilowatt hour energy potential for all conventional nonrenewable energy resources was compiled by superimposing, one upon the other, the separate maps for the energy potential of each energy resource. The resulting information was analyzed and regions of varying intensity in energy potential were formulated. Boundaries for these regions, of necessity, represent an intellectual generalization rather than a definitive mathematical computation. Criteria established for drawing these boundaries were elementary: they must separate and encompass regions with distinctly different energy potentials; and the boundaries were to reflect the traditionally established regions of the U.S.S.R. in order to facilitate comparison with productive data.

Fourteen distinct energy potential regions were formulated (see Fig. 19). The regional pattern of kilowatt hour energy potential of all nonrenewable energy resources is dominated by coal. This does not obviate the importance of the other energy resources no matter how

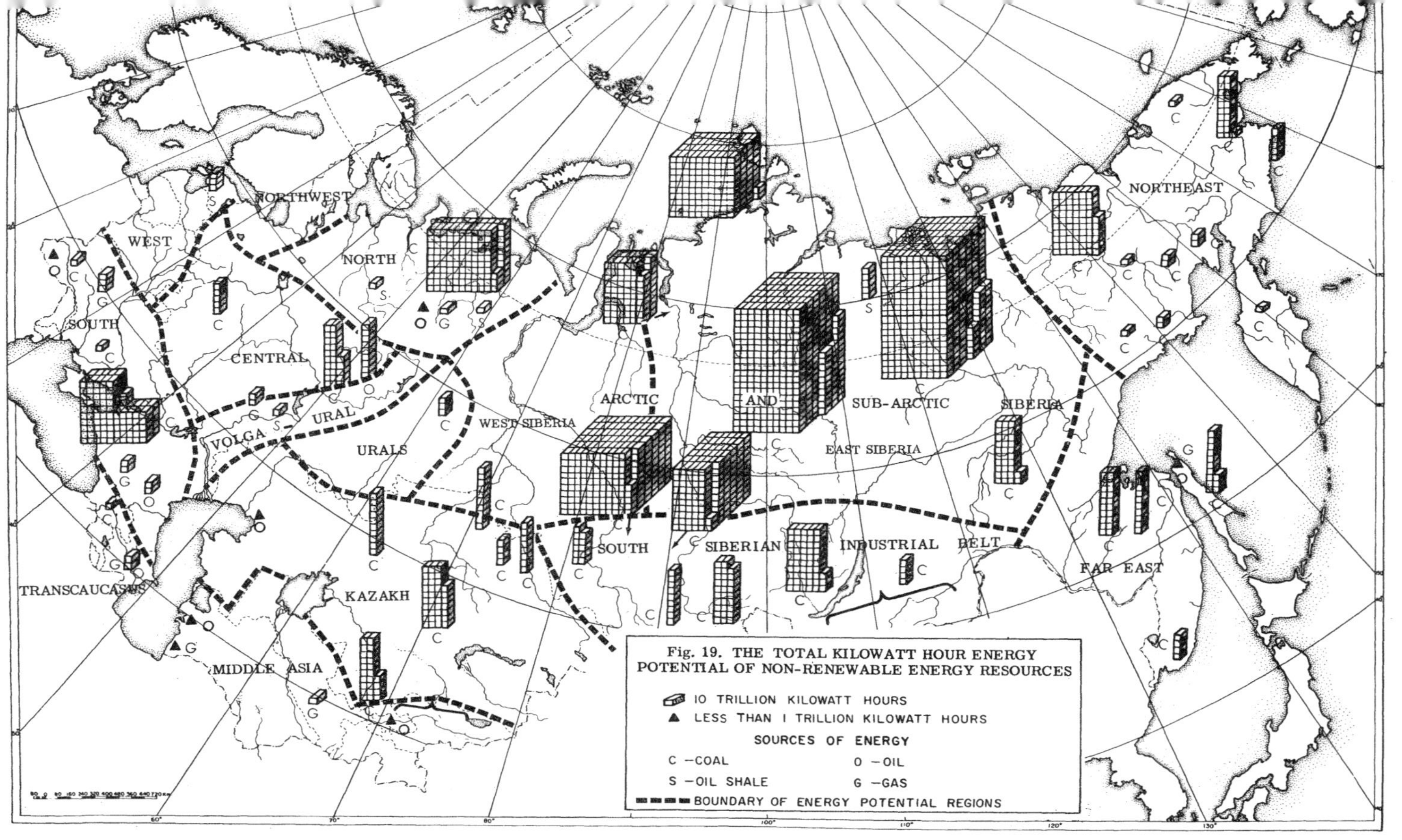

Fig. 19. THE TOTAL KILOWATT HOUR ENERGY POTENTIAL OF NON-RENEWABLE ENERGY RESOURCES

minute they may appear to be in the total energy balance. Their significance depends on their location in the regional energy resource base.

SYMPOSIUM ON THE REGIONAL ENERGY RESOURCE BASE OF THE U.S.S.R.

European Russia, including the Urals and the Caucasus, contains eight of the 14 energy potential regions on the montage map. Collectively, these regions contain approximately 9.0 per cent of the total kilowatt hour energy potential of the nonrenewable reserves, with coal accounting for approximately 8.4 per cent of this potential. The vast territory east of the Urals and north of Kazakhstan and Middle Asia contains five of these regions, which have approximately 89 per cent of the total energy potential. Coal is the one dominating energy resource in these regions. Kazakhstan and Middle Asia contain the remaining two per cent of the energy potential.

According to computations based on reserve data compiled from Soviet sources, there are two regions without a nonrenewable energy potential base. Reportedly, the northwestern region of the European section is devoid of conventional nonrenewable energy resources. Western Siberia, as delimited in this study, is also without proven nonrenewable resources. Soviet speculation may attribute possibilities to the region, but at this writing, definitive reports are unknown.

European Russia contains three regions characterized by their possession of a single energy resource potential. The West has but one nonrenewable energy resource, shale. The Central and Urals Regions contain only coal, primarily brown coal.

Two adjacent regions in European Russia contain three nonrenewable energy resources, coal, oil, and gas. The South, which includes the Ukraine, Moldavia, and the North Caucasus, is one of the larger of the two regions and is important because of the Donets coal basin. The Transcaucasus Region is based on the oil and gas of Azerbaidzhan and the coal and oil of Georgia.

European Russia's remaining two regions, the North and the Volga-Pre Ural, possess all four of the conventional nonrenewable energy resources. Coal in the Pechora Basin dominates the energy potential of the North but it is augmented by oil shale, gas, and oil. Highly sulfurous coal from the unexploited Kama Basin accounts for approximately half of the energy potential of the Volga-Pre Ural Basin. Oil and gas, supplemented by a small quantity of shale, account for the remaining energy potential.

The Dominance of Eastern Regions.—Coal based energy potential regions east of the Urals and north of Kazakhstan possess approximately 89 per cent of the kilowatt hour energy potential of the U.S.S.R. Vast and vacant Arctic and Subarctic Siberia is the major energy potential region, with the hugh Tanguska and Lena coal basins, augmented by the Northeast Shale Basin. Northeastern Siberia derives its energy potential from one nonrenewable energy resource, coal. The Soviet Far East is the only energy potential region east of the Urals possessing three of the nonrenewable energy resources. Oil and incidental oil gas supplement coal which dominates this energy potential region, as well as all others east of the Urals. The South Siberian Belt and Transbaikalia form one energy potential region, based solely on coal.

Kazakhstan and Middle Asia.—The two remaining energy potential regions of Kazakhstan and Middle Asia possess 2.0 per cent of the kilowatt hour energy potential of the nonrenewable resources. Kazakhstan possesses about 1.5 per cent of the Soviet energy potential and Middle Asia approximately 0.5 per cent. Coal, natural gas, and oil constitute the energy potential base of these regions.

POTENTIAL AND DEVELOPED REGIONS COMPARED

A comparison of the potential and developed energy resource base, emanating from computations on Soviet reports of reserves and production, reveals that the major energy potential regions lie east of the Urals and are coal based. Soviet sources state that they were relatively undeveloped through 1955, even though they contain approximately 89 per cent of the energy potential. In contrast, European Russia's energy potential regions were intensively developed. European Russia possesses 9 per cent of the kilowatt hour energy potential of the Soviet Union. It contains only 8.4 per cent of the kilowatt hour energy potential of coal but produced 65 per cent of the coal in 1955. Approximately 57 per cent of the energy potential of oil shale is located in this section; yet in 1955, it produced 100 per cent of the shale. About 92 per cent of the energy potential of Soviet oil reserves are located here, and in 1955 it produced approximately 91 per cent of the oil. European Russia also possesses 87 per cent of the natural gas. In 1955, it produced 95 per cent of the total natural gas in the country.

Although the energy potential regions of European Russia contain the greatest variety of nonrenewable energy resources, cumulatively these resources account for little of the total Soviet energy potential. However, this fact should not depreciate the practical significance of these resources. Since oil shale is the only nonrenewable energy resource in the Western region, its development achieves importance. The same is true of brown coal in the Central Region. The Volga-Pre Ural Region has developed deposits of oil and gas and some oil shale; yet these resources constitute less than half of the energy potential of the region. One coal deposit, the Kama, accounts for more than half of the energy potential of the region, but its quality does not make exploitation feasible.

Two regions, the South and the North, possess approximately 90 per cent of the kilowatt hour energy potential in European Russia. About 60 per cent of this potential is in the North, and the remaining 40 per cent in the South. In addition to coal, which is the principal energy resource, they both contain significant deposits of oil and gas. Development of the two regions offers sharp contrasts. The South, with its famed Donets Basin, is highly developed and plans call for a concerted effort to develop oil and gas deposits. The North, with its extensive Pechora Basin and Komi-Ukhta oil and gas fields, is just now emerging as a developed energy potential region. Geographical location and historical antecedents have been the deciding factors in the contrasting development.

Energy potential of the Urals Region totals less than one-tenth of one per cent, but this single energy resource region, based on coal alone, is highly developed. While it imported large quantities of coal in 1955, it also exported some coal to the Volga-Pre Ural Region.

The most outstanding characteristic concerning the energy potential of European Russia is the fact that regions with comparatively low energy potential, such as the West, the Central Region, and the Urals, are centers of industrial concentration. A second important characteristic is the historically significant role the South and the Transcaucasus have had in supplying energy for the development of the rest of European Russia. A third characteristic is the great potential possessed by the North and Volga-Pre Ural Regions and their possible influence on the future development of European Russia.

The Eastern Regions.—According to computations based on Soviet reports of energy reserves, regions east of the Urals and north of Kazakhstan have approximately 89 per cent of the kilowatt hour energy potential of the Soviet Union. Coal accounts for all but an infinitesimal amount of this energy. The largest potential and the least development occur in Arctic and Subarctic Siberia. Remote and isolated, this region is the largest single energy potential region in the U.S.S.R. The equally isolated and remote Northeast has approximately two per cent of the energy potential of the Soviet Union; except for local development, it remains unexploited.

The South Siberian Belt is the second largest of the energy potential regions. Based on a single energy resource, coal, it has approximately 23 per cent of the total energy potential of the Soviet Union.

Among eastern regions, it is the most highly developed. Previously cited plans indicate that development is to be intensified in the next 10 to 15 years.

Far Eastern potentials are only half of one per cent. The only multiresource energy potential region east of the Urals, development here has long been an established fact.

Energy potential regions of the East have one outstanding energy resource, coal, which dominates the entire energy potential of the Soviet Union. Their existence is a reflection of the distribution of this resource. According to Soviet reports, most of the energy potential in this region is the result of recent discovery. Developmental plans for the future, however, must take into account the rather difficult environmental conditions.

CONCLUSIONS

When all conventional nonrenewable energy resources are measured in terms of a single energy criterion, kilowatt hours, a regional pattern for these resources evolves.

This study is dominated by the preponderance of coal in the energy resource base of the U.S.S.R. and, in a sense, reflects the regional distribution of coal.

Energy potentials and reserves of oil, natural gas, and oil shale are located primarily in European Russia, including the Urals and Caucasus. Similarly, production and consumption of all types of energy resources are centered in this European section. However, within European Russia itself, most of the consumption takes place in the South, the Central Regions, and the Urals, and, in order to satisfy the needs of these consumers, large quantities of coal are imported from basins east of the Urals. In order to minimize the importation of coal from remote basins, the local fuels, such as oil shale and brown coal, are utilized to the fullest extent in the European Basin.

Significantly, the preponderance of Soviet energy lies east of the Urals in the form of a single energy resource, coal. Here, too, according to Soviet data, mining costs are reportedly less expensive. These are the factors that may have influenced Soviet planners in their decision to emphasize economic development in the eastern regions over the next decade.

GENERAL APPENDIX

TABLE I

GROUPING OF ELECTRIC STATIONS BY TYPE OF ENERGY RESOURCE EACH CONSUMED IN 1955

Type of Station	Capacity of Station Kwt. 10^3 [a]	Production of Electro Energy Kwt. Hrs. 10^3 [a]	Per Cent of Total Production [b]	Kwt. Hours Per Person [c]
All Electric Stations (stationary and mobile)	36,236	170,225	100	850.2
Hydro Station	5,986	23,165	13.61	115.8
Oil Fuel	6,233	15,290	8.98	76.5
Coal	20,473	109,139	64.12	542.7
Peat	2,234	12,422	7.29	64.1
Shale	274	1,314	0.77	6.6
Wood	528	1,091	0.64	5.5
Natural Gas	591	3,036	1.80	15.3
Generated Gas	67	77	0.05	0.4
Other Gas Generated Fuel	541	3,748	2.20	18.7
Other Energy Resources	309	913	.54	4.6

[a] Tsentralnoe Statisticheskoe Upravlenie pri Sovete Ministrov S.S.S.R., (K. G. Ivanova, ed.), *Promyshlennost SSSR, Statisticheskiy Sbornik* (Moskva: Gosstatizdat, 1957), p. 178.

[b] Computed from Column 2.

[c] Statisticheskoe Upravlenie R.S.F.S.R., *Narodnoe Khozyaistvo RSFSR, Statisticheskoe Sbornik* (Moskva: Gosstatizdat, 1957), p. 17; and Ivanova, ed., *ibid.*, p. 178.

COAL APPENDIX

TABLE I

TOTAL GEOLOGICAL RESERVES OF COAL IN THE U.S.S.R., AS OF 1 JANUARY 1956 IN BILLIONS OF METRIC TONS (10^9)

Basins and Deposits	Totals	By Degree of Reliability		
		Proven	Prob-able	Possible
In the entire U.S.S.R.	8,669.51	241.21	941.89	7,486.41
I. European Part of the U.S.S.R.	647.28	74.91	105.63	466.74
Donets Basin	240.62	57.16	79.53	103.93
incl. Ukraine SSR	173.16	49.00	56.49	67.67
Rostov and Kamen Oblasts R.S.F.S.R.	67.46	8.16	23.04	36.26
Lvov-Volyn coal bearing region	1.75	1.65	0.10	---
Dneper basin	4.18	3.05	0.99	0.14
Moscow basin	24.31	8.89	5.32	10.10
Kama basin	30.32	---	---	30.32
Pechora basin	344.50	4.10	19.40	321.00
Other deposits in the European Part of U.S.S.R.	1.60	0.06	0.29	1.25
II. Caucasus	2.01	0.56	0.43	1.02
Tkibuli deposits	0.46	0.29	0.17	---
Tkvarchal deposits	0.08	0.07	0.01	---
Akhaltsikh deposits	0.15	0.10	---	0.05
Deposits of the Northern Slope of the Caucasus	1.32	0.10	0.25	0.97
III. Urals	7.51	5.00	1.56	0.95
Kizel basin	1.06	0.61	0.32	0.13
South Ural basin	1.76	1.56	0.20	---
Chelyabinsk basin	1.63	1.37	0.19	0.07
Orsk (East Ural) basin	1.03	0.62	0.41	---
North-Sosvinsk coal bearing region	1.07	0.21	0.29	0.57
Other deposits of the Urals	0.96	0.63	0.15	0.18

By Depth Zones (In Meters)				By Degree of Metamorphism (In Terms of the Classification System for Donets Grades of Coal)					
0 - 300	300 - 600	600 - 1200	1200 - 1800	A & T	PS, K, PZH	G	D	DB	B
2,351.51	1,779.88	2,838.03	1,700.09	1,080.45	1,952.71	819.32	1,330.02	474.09	3,012.91
115.44	132.80	237.09	161.95	124.28	149.47	169.99	121.63	51.82	30.09
28.39	50.05	96.41	65.77	93.48	32.46	65.05	28.13	21.50	---
20.90	35.53	71.05	45.68	39.78	28.63	63.57	19.75	21.43	---
7.49	14.52	25.36	20.09	53.70	3.45	1.86	8.45	---	---
0.06	1.55	0.14	---	---	0.01	1.74	---	---	---
4.18	---	---	---	---	---	---	---	---	4.18
24.31	---	---	---	---	---	---	---	---	---
---	---	24.64	5.68	---	---	---	---	30.32	---
57.00	81.10	115.90	90.50	30.80	117.00	103.20	93.50	---	---
0.10	---	---	---	---	---	---	---	---	---
0.78	1.14	0.09	---	0.01	0.12	1.73	---	---	0.15
0.24	0.17	0.05	---	---	---	0.46	---	---	---
0.08	---	---	---	---	0.08	---	---	---	---
0.04	0.07	0.04	---	---	---	---	---	---	0.15
0.42	0.90	---	---	0.01	0.04	1.27	---	---	---
5.95	0.82	0.62	0.12	0.31	1.06	0.23	---	---	5.91
0.32	0.20	0.43	0.11	---	1.06	---	---	---	---
1.76	---	---	---	---	---	---	---	---	1.76
1.04	0.45	0.13	0.01	---	---	---	---	---	1.63
1.03	---	---	---	---	---	---	---	---	1.03
1.07	---	---	---	---	---	---	---	---	1.07
0.73	0.17	0.06	---	0.31	---	0.23	---	---	0.42

TABLE I (Continued)

Basins and Deposits	Totals	By Degree of Reliability		
		Proven	Prob-able	Possible
IV. Western and Eastern Siberia (Southern Part)	2,290.05	114.49	554.77	1,620.79
Gorlov basin	17.23	0.07	2.26	14.90
Kuznetsk basin	905.30	70.88	253.51	580.91
Deposits of Tomsk Oblast	2.59	---	---	2.59
Kansk-Achinsk basin	1,220.30	35.00	234.28	951.02
Minusinsk basin	36.94	2.31	32.98	1.65
Deposits of the Tuvinian Autonomous Oblast	18.68	1.06	2.20	15.42
Irkutsk basin	88.90	5.17	29.54	54.19
Other deposits Western Siberia	0.11	---	---	0.11
V. Eastern Siberia (Northern Part)	5,237.87	4.11	190.65	5,043.11
Taimyr basin	583.50	0.25	23.25	560.00
Tunguska basin	1,744.77	1.40	53.37	1,690.00
Ust-Yenisey coal bearing region	221.75	---	4.75	217.00
South-Yakutsk coal bearing region	40.05	0.65	2.40	37.00
Other deposits of Eastern Siberia	0.56	0.02	0.01	0.53
Lena basin	2,647.24	1.79	106.87	2,538.58
VI. The Northeast	239.97	0.85	9.28	229.84
Zyryansk coal area	102.60	0.34	2.26	100.00
Omsukchansk coal area	2.87	0.01	0.03	2.83
Omolonsk coal area	0.55	---	---	0.55
Chaun-Chukotsk coal area	1.30	---	---	1.30
Anadyr coal area	97.71	0.01	---	97.70
Elgen deposit	2.93	0.03	0.50	2.40
Avekov deposit	13.91	0.01	---	13.90
Okhotsk coal area	10.00	0.20	3.30	6.50
Arkagalin coal area	1.09	0.15	0.01	0.93
Bukhta-Ugolnaya	6.18	0.07	2.38	3.73
Kamchatka Peninsula (all)	0.83	0.03	0.80	---
VII. Transbaikal, Far East, Sakhalin	64.14	8.74	16.72	38.68
Transbaikal deposits	8.37	1.63	1.43	5.31
Bureya basin	25.02	1.08	9.24	14.70
Bikinsk deposit	2.90	1.43	0.51	0.96
Suifun basin	1.66	0.37	0.15	1.14
Suchan basin	1.43	0.19	0.34	0.90
Uglov basin	1.02	0.65	0.26	0.11
Maikhinsk deposit	0.60	0.32	0.28	---
Suputinsk deposit	0.45	0.10	0.35	---
Sakhalin deposits	20.09	2.01	3.88	14.20
Other deposits of the Far East	2.60	0.96	0.28	1.36

* The grade of this coal is not represented in this analysis.

By Depth Zones (In Meters)				By Degree of Metamorphism (In Terms of the Classification System for Donets Grades of Coal)					
0-300	300-600	600-1200	1200-1800	A & T	PS, K, PZH	G	D	DB	B
520.30	301.82	810.42	657.51	280.65	223.27	361.62	143.17	55.90	1,225.44
3.30	3.80	6.53	3.60	17.23	---	---	---	---	---
156.48	169.34	302.36	277.12	263.42	216.00	304.82	65.16	55.90	---
2.59	---	---	---	---	---	---	---	---	2.59
249.90	114.70	487.30	368.40	---	---	1.70	0.06	---	1,218.54
17.50	8.50	9.29	1.65	---	---	7.51	29.43	---	---
3.25	3.75	4.94	6.74	---	7.27	11.14	---	---	---
87.17	1.73	---	---	---	---	36.20	48.50	---	4.20
0.11	---	---	---	---	---	---	---	---	0.11
1,514.40	1,228.82	1,670.25	824.40	666.59	1,384.15	247.70	1,047.80	357.52	1,534.11
83.30	104.30	242.90	153.00	143.10	412.30	---	---	---	28.10
575.00	485.77	646.00	38.00	522.00	737.00	52.00	243.00	190.77	---
51.50	63.40	67.85	39.00	---	---	---	55.00	166.75	---
24.95	15.10	---	---	---	39.60	0.45	---	---	---
0.56	---	---	---	---	---	---	---	---	0.56
779.09	560.25	713.50	594.40	1.49	390.50*		749.80		1,505.45
85.40	58.48	69.17	26.92	2.87	106.99	3.84	1.72	---	124.55
34.30	24.40	31.30	12.60	---	102.60	---	---	---	---
0.89	0.79	1.07	0.12	2.87	---	---	---	---	---
0.52	0.03	---	---	---	---	0.55	---	---	---
0.10	0.20	0.60	0.40	---	1.30	---	---	---	---
24.51	24.10	35.40	13.70	---	---	---	---	---	97.71
2.93	---	---	---	---	---	---	---	---	2.93
6.96	6.95	---	---	---	---	---	---	---	13.91
10.00	---	---	---	---	---	---	---	---	10.00
0.88	0.21	---	---	---	---	---	1.09	---	---
3.48	1.80	0.80	0.10	---	6.18*		---	---	---
0.83	---	---	---	---	---	0.20	0.63	---	---
30.62	16.28	11.43	5.81	0.97	2.25	27.34	9.41	8.85	15.32
6.72	1.53	0.12	---	---	---	1.18*		3.80*	3.39
6.19	7.20	7.63	4.00	0.02	---	25.00	---	---	---
2.71	0.19	---	---	---	---	---	---	---	2.90
0.75	0.85	0.06	---	---	0.03	---	1.63	---	---
0.29	0.24	0.59	0.31	0.34	0.32	0.66	0.11	---	---
0.51	0.38	0.13	---	---	---	---	---	0.11	0.91
0.50	0.10	---	---	---	---	---	---	---	0.60
0.35	0.10	---	---	---	---	---	---	---	0.45
10.38	5.33	2.88	1.50	0.05	1.80	0.85	7.08	4.94	5.37
2.22	0.36	0.02	---	0.56	0.10	0.24	---	---	1.70

TABLE I (Continued)

Basins and Deposits	Totals	By Degree of Reliability		
		Proven	Probable	Possible
VIII. Kazakhstan	139.90	28.99	57.23	53.68
Karaganda basin	51.23	10.30	25.76	15.17
Ubagani basin	36.49	6.32	10.53	19.64
Ekibastuz deposit	12.21	9.11	1.30	1.80
Lenger deposit	2.02	0.12	0.20	1.70
Other Kazakh deposits	16.94	1.81	6.32	8.80
Maikyuben	21.01	1.33	13.11	6.57
IX. Middle Asia	40.78	3.55	5.62	31.61
Hissar coal region	3.78	0.06	0.11	3.61
South Hissar region	1.67	0.01	0.01	1.65
South Tadzhik depression	2.46	---	---	2.46
Ziddin deposit	1.44	0.02	0.07	1.35
Magian deposit	1.07	0.01	0.04	1.02
Fan-Yagnob deposit	1.78	0.34	0.37	1.07
Other deposits of the Zeravshan region	1.93	---	---	1.93
Nazar-Ailoksk deposit	0.44	---	0.02	0.42
Shuroabad-Ravnous deposit	0.68	0.02	0.04	0.62
Mionadus deposit	1.47	---	0.03	1.44
Sulyukta deposit	0.76	0.17	0.16	0.43
Angren deposit	2.82	1.52	0.40	0.90
Tashkent region	0.64	---	---	0.64
Shurab deposit	3.08	0.13	1.14	1.81
Kyzyl-Kiya deposit	2.38	0.11	0.17	2.10
North Fergana deposit	2.55	0.15	0.47	1.93
Aldyyar deposit	0.48	0.02	0.13	0.33
Kok-Yangak deposits	2.07	0.22	0.08	1.77
East Fergana (Uzgen) basin	3.09	0.27	1.66	1.16
Minkush coal region	4.21	0.47	0.70	3.04
Other deposits of Middle Asia	1.98	0.03	0.26	1.69

Source: *Zapasy Uglei i Goryuchikh Slantsev SSSR*, ed. by N. V. Shabarova and A. V. Tyzhnova, Moscow: Gosgeoltekhizdat, 1958, pp. 12–16.

See also: A. Zademikdo, "Rezervy Ekonomicheskikh Raionov – Na Sluzhby Rodine," (Reserves of the Economic Region – in the Service of the Fatherland) *Pravda*, 22 July, 1957. No. 203 (14232)

By Depth Zones (In Meters)				By Degree of Metamorphism (In Terms of the Classification System for Donets Grades of Coal)					
0-300	300-600	600-1200	1200-1800	A & T	PS, K, PZH	G	D	DB	B
69.04	30.54	26.13	14.20	1.83	72.69	2.18	0.04	---	63.16
9.74	9.04	19.64	12.81	---	49.73	0.63	---	---	1.17
29.83	6.66	---	---	---	---	---	---	---	36.49
4.62	5.84	1.75	---	---	12.16	0.05	---	---	---
0.08	0.35	1.11	0.48	---	---	---	---	---	2.02
10.75	2.29	3.00	0.90	1.83	11.10	1.50	0.04	---	2.47
14.02	6.36	0.63	---	---	---	---	---	---	21.01
9.58	9.18	12.83	9.18	2.94	12.77	4.64	6.25	---	14.18
0.33	0.72	1.37	1.36	0.45	3.27	0.06	---	---	---
0.34	0.35	0.70	0.28	---	1.67	---	---	---	---
---	---	1.23	1.23	---	2.46	---	---	---	---
1.44	---	---	---	---	1.44	---	---	---	---
0.15	0.13	0.31	0.48	---	---	---	1.07	---	---
0.75	0.32	0.48	0.23	---	0.98	0.80	---	---	---
0.72	1.15	0.06	---	---	---	1.93	---	---	---
0.28	0.11	0.05	---	0.44	---	---	---	---	---
0.12	0.14	0.17	0.25	---	0.68	---	---	---	---
0.24	0.25	0.48	0.50	0.67	0.80	---	---	---	---
0.11	0.23	0.22	0.20	---	---	---	---	---	0.76
0.95	0.98	0.73	0.16	---	---	---	---	---	2.82
---	---	0.28	0.36	---	---	---	---	---	0.64
0.39	0.71	1.24	0.74	---	---	---	---	---	3.08
0.07	0.55	1.21	0.55	---	---	---	---	---	2.38
0.57	0.65	0.75	0.57	---	---	---	2.55	---	---
0.12	0.07	0.11	0.18	---	---	0.48	---	---	---
0.57	0.68	0.53	0.29	---	---	---	2.07	---	---
1.17	0.67	0.85	0.40	1.03	1.02	1.04	---	---	---
1.01	0.92	1.27	1.01	---	---	---	---	---	4.21
0.25	0.55	0.79	0.39	0.35	0.45	0.33	0.56	---	0.29

p. 2. – P. Antropov, "Prirodnye Bogatstva Strany – Na Sluzhby Narodnomu Khozyaistbu" (Natural Resources of the Country – in the Service of the National Economy) *Pravda*, 4 Dec., 1957, No. 338 (14367) p. 2-3. – Anon., "Towards the Creation of a Ferrous Metallurgical Base in Southern Yakutsk," *Geografiya V. Shkole*, Moskva, No. 3, 1958, p. 62. – V. Vityazeva, "Concerning the Development of the Pechora Coal Basin," *Planovoe Khozyaistvo*, No. 11, 1957, p. 79.

TABLE II

ENERGY POTENTIALS OF THE MINEABLE RESERVES OF SOVIET COAL BASINS AND DEPOSITS FOR JANUARY 1, 1956

Basins and Deposits by Type of Coal	Mineable Reserves in 10^9 Tons (1)	Calorie Content Mineable Reserves in Cal. 10^{12} (2)	Kwt. Hour Content Mineable Reserves in Kwt. Hrs. 10^6 (3)
Total U.S.S.R.	7,765.29	41,378,110	48,233,118,420
Hard Coal	4,903.53	34,324,710	39,912,772,788
Brown Coal	2,861.76	7,154,400	8,320,345,632
I. European Part	506.55	3,444,195	4,004,909,146
Hard Coal	483.96	3,387,720	3,939,240,816
Brown Coal	22.59	56,472	65,669,130
Donets Basin (Hard)	189.97	1,329,790	1,546,279,812
Ukraine SSR	146.61	1,026,270	1,193,346,756
Rostov and Kamen Oblasts	43.36	303,520	352,933,056
Lvov-Volyn Coal Region (Hard)	1.42	9,940	11,558,232
Dneper Basin Brown Coal	3.66	9,150	10,639,620
Moscow Basin Brown Coal	17.48	43,700	50,814,360
Kama Basin (Hard)	30.20	211,400	245,815,920
Pechora Basin (Hard)	262.37	1,836,590	2,135,586,852
Other Deposits of the European Part Brown Coal	1.45	3,625	4,125,150
II. Caucasus	1.01	6,485	7,540,758
Hard Coal	0.88	6,160	7,162,848
Brown Coal	0.13	352	377,910
Tkibuli Deposit Hard Coal	0.46	3,220	3,744,216
Tkvarchal Deposit Hard Coal	0.08	560	651,168
Akhaltsikh Deposit Brown Coal	0.13	325	377,910
Deposits of the North Slope of the Caucasus (Hard)	0.34	2,380	2,767,464

(1) Mineable reserves of coal are based on the criteria of depth and thickness of coal seams, and the ash content of the coal. Geographical location, transportation facilities, and hydrogeological conditions were not considered. Elements of the criteria may vary with the individual coal deposits. See *Zapasy Uglei i Goruchikh Slantsev SSSR* (Moskva, Gosgeoltekhizdat, 1958), p. 30.

(2) Soviet standards for the conversion of hard coal and brown coal to their calorie equivalents were used in this computation. See p. 68 of Yu. Vasilev and K. Pospelova, "Strukturnye Sdvigi V Toplevnom Balanse S Sh. A.," *Planovoe Khozyaistvo*, No. 6 (1957), pp. 66-79. Hard coal – K/Kal. 7 mil. to a metric ton, brown coal – 2.5 mil K/Kal. to a metric ton.

(3) The factor used in converting calories to Kwt. hours was × 0.0011628. See p. 1,720, Norbert A. Lange, Ph.D., Ed., *Handbook of Chemistry* (Sandusky: Handbook Publishers Inc., 1944).

TABLE II (Continued)

Basins and Deposits by Type of Coal	Mineable Reserves in 10^9 Tons (1)	Calorie Content Mineable Reserves in Cal. 10^{12} (2)	Kwt. Hour Content Mineable Reserves in Kwt. Hrs. 10^6 (3)
III. Urals	6.95	24,395	28,366,506
Hard Coal	1.56	10,920	12,697,776
Brown Coal	5.39	13,475	15,668,730
Kizel Basin (Hard)	1.06	7,420	8,627,976
South Ural Basin Brown Coal	1.46	3,650	4,244,220
Chelyabinsk Basin Brown Coal	1.51	3,775	4,389,570
Orsk Basin – East Ural Brown Coal	0.99	2,475	2,877,930
North Sosvinsk Coal Region – Brown Coal	1.55	2,625	3,052,350
Other Deposits of Urals	0.88	4,450	5,174,460
Hard Coal	0.50	3,500	4,069,800
Brown Coal	0.38	950	1,104,660
IV. Western Siberia and So. Part of Eastern Siberia	2,144.30	9,544,480	11,109,949,344
Hard Coal	913.94	6,522,580	7,584,456,024
Brown Coal	1,212.36	3,031,900	3,525,493,320
Gorlov Basin (Hard)	15.06	105,420	122,582,376
Kuznetsk Basin Hard Coal	804.17	5,629,190	6,545,622,132
Deposits of Tomsk Oblast – Brown Coal	2.59	6,475	7,529,130
Kansk-Achinsk Basin	1,207.79	3,027,630	3,520,528,164
Hard Coal	1.59	11,130	12,941,964
Brown Coal	1,206.20	3,016,500	3,507,586,200
Minusinsk Basin Hard Coal	36.30	254,100	295,467,480
Deposits of the Tuvinian A. O. (Hard)	10.83	75,810	88,151,868
Irkutsk Basin	67.35	455,580	529,748,424
Hard Coal	63.89	446,930	519,690,204
Brown Coal	3.46	8,650	10,058,220
Other Deposits of Western Siberia Brown Coal	0.11	275	319,770
V. Eastern Siberia Northern Part	4,699.85	26,378,845	30,674,483,756
Hard Coal	3,251.16	22,728,120	26,463,141,936
Brown Coal	1,448.69	3,620,725	4,211,341,820

TABLE II (Continued)

Basins and Deposits by Type of Coal	Mineable Reserves in 10^9 Tons (1)	Calorie Content Mineable Reserves in Cal. 10^{12} (2)	Kwt. Hour Content Mineable Reserves in Kwt. Hrs. 10^6 (3)
Taimyr Basin	511.58	3,454,430	4,016,811,204
Hard Coal	483.44	3,384,080	3,935,008,224
Brown Coal	28.14*	70,350	81,802,980
Tanguska Basin			
Hard Coal	1,516.02	10,612,140	12,339,796,392
Ust-Yenisey Coal Region (Hard)	214.16	1,499,120	1,743,176,736
Lena Basin	2,417,98	10,535,410	12,250,574,748
Hard Coal	997.88	6,985,160	8,122,344,048
Brown Coal	1,420.10	3,550,250	4,128,230,700
South Yakutsk			
Hard Coal	39.66	277,620	322,861,536
Other Deposits of Eastern Siberia			
Brown Coal	0.45	1,125	1,308,150
VI. Northeast U.S.S.R.	183.85	916,105	1,023,386,094
Hard Coal	93.44	654,080	760,564,224
Brown Coal	90.41	226,025	262,821,870
Zyryansk Coal Area			
Hard Coal	81.26	568,820	661,423,896
Omsukchan Coal Area			
Hard Coal	2.60	18,200	21,162,960
Omolonsk Coal Area	0.55	3,850	4,476,780
Chun-Chukotsk Coal Area (Hard Coal)	1.30	9,100	10,581,480
Anadry Coal Area			
Brown Coal	70.78	176,950	205,757,460
Elgen Deposit			
Brown Coal	2.86	7,150	8,314,020
Avekovsk Deposit			
Brown Coal	8.40	21,000	24,418,800
Okhotsk Coal Area			
Brown Coal	8.37	20,925	24,331,590
Arkagalin Coal Area			
Hard Coal	0.72	5,040	3,860,512
Bukhta Coal (Hard)	6.18	43,260	50,302,728
Kamchatka Deposits	0.83	5,810	6,755,868

* In this instance the balance of reserves exceeds the Geological Reserves which is an obvious impossibility.

TABLE II (Continued)

Basins and Deposits by Type of Coal	Mineable Reserves in 10^9 Tons (1)	Calorie Content Mineable Reserves in Cal. 10^{12} (2)	Kwt. Hour Content Mineable Reserves in Kwt. Hrs. 10^6 (3)
VII. Transbaikal, F.E. and Sakhalin	61.96	368,895	426,625,506
Hard Coal	47.11	329,770	383,456,556
Brown Coal	14.85	39,125	43,168,950
Transbaikal Deposits	7.25	36,395	42,320,106
Hard Coal	4.06	28,420	33,046,776
Brown Coal	3.19	7,975	9,273,330
Bureya Basin (Hard)	24.94	174,580	203,001,624
Bikinsk Deposit Brown Coal	2.90	7,250	8,430,300
Suifun Basin (Hard Coal)	1.62	11,340	13,186,152
Suchan Basin (Hard)	1.40	9,800	11,395,440
Uglov Basin (Artem)	0.85	2,575	2,994,210
Hard Coal	0.10	700	813,960
Brown Coal	0.75	1,875	2,180,250
Maikhinsk Deposit Brown Coal	0.59	1,475	1,715,130
Suputinsk Deposit Brown Coal	0.46	1,150	1,337,220
Sakhalin Island	19.41	102,330	130,152,204
Hard Coal	14.09	98,630	114,686,964
Brown Coal	5.32	13,300	15,465,240
Other Deposits of the Far East	2.53	10,330	12,011,724
Hard Coal	0.89	6,230	7,244,244
Brown Coal	1.64	4,100	4,767,480
VIII. Kazakhstan	122.53	616,375	716,720,850
Hard Coal	68.90	482,300	560,818,400
Brown Coal	53.63	134,075	155,902,410
Karaganda Basin	46.55	321,485	373,822,758
Hard Coal	45.58	319,060	371,002,968
Brown Coal	0.97	2,425	2,819,790
Ubagan Basin Brown Coal	35.40	88,500	102,907,800
Ekibastuz Deposit Hard Coal	10.76	75,320	87,582,096
Maikyuben Deposit (Brown)	13.42	33,550	39,011,940
Lenger Deposit Brown Coal	1.98	4,950	5,755,860
Other Deposits of Kazakhstan	14.42	92,570	107,640,396
Hard Coal	12.56	87,920	102,233,376
Brown Coal	1.86	4,650	5,407,020

TABLE II (Continued)

Basin and Deposit by Type of Coal	Mineable Reserves in 10^9 Tons (1)	Calorie Content Mineable Reserves in Cal. 10^{12} (2)	Kwt. Hour Content Mineable Reserves in Kwt. Hrs. 10^6 (3)
IX. Middle Asia	38.29	206,335	239,926,338
Hard Coal	24.58	172,060	200,071,368
Brown Coal	13.71	34,275	39,854,970
Hissar Coal Region Hard Coal	1.96	13,720	15,953,616
South Hissar Coal Region (Hard)	1.67	11,690	13,593,132
South Tadzhik Depression (Hard)	2.46	17,220	20,023,416
Ziddinsk Deposit Hard Coal	1.44	10,080	11,721,024
Magiansk Deposit Hard Coal	1.06	7,420	8,627,976
Fan-Yagnob Deposit (Hard)	1.77	12,390	14,407,092
Zervashan Region Hard Coal	1.93	13,510	15,709,428
Nazar-Ailoksk Deposit (Hard)	0.44	3,080	3,581,524
Shuroabad-Ravnous Deposit (Hard)	0.68	4,760	5,534,928
Mionadus Deposit Hard Coal	1.47	10,290	11,965,212
Sulyuktia Deposit Brown Coal	0.68	1,700	1,976,760
Angren Deposit Brown Coal	2.80	7,000	8,139,600
Tashkent Region Brown Coal	0.64	1,600	1,860,480
Shurab Deposit Brown Coal	2.82	7,050	8,197,740
Kyzyl-Diisk Deposit Brown Coal	2.27	5,675	6,597,800
North Fergana Deposit Hard Coal	2.54	17,780	20,674,584
Aldyyar Deposit Hard Coal	0.47	3,290	3,825,612
Kok-Yangak Deposit Hard Coal	2.07	14,490	16,848,972
East Fergana Uzgen Deposit (Hard)	3.01	21,070	24,500,196
Minkushsk Coal Region (Brown)	4.21	10,525	12,238,470
Other Deposits M.A.	1.90	11,995	13,947,786
Hard Coal	1.61	11,270	13,104,756
Brown Coal	0.29	725	843,030

TABLE III

CHARACTERISTICS OF SOVIET COALS

Basins and Deposits by Type of Coal	Moisture Content of Coal Total %	Moisture Content of Coal Dry %	Ash Content Absolutely Dry Coal %
EUROPEAN RUSSIA			
Donets Basin – D	9.5 – 21.0	3.1 – 7.5	6.0 – 27.0
G	3.0 – 12.0	1.0 – 4.8	3.0 – 32.0
Pzh	3.0 – 8.0	0.4 – 2.1	2.0 – 31.0
K	3.0 – 12.0	1	5.0 – 21.0
PS	3.0 – 6.5	0.8	5.0 – 20.0
T	3.0 – 6.5	0.4 – 1.8	4.3 – 25.0
A	3.5 – 9.0	2	7.0 – 19.0
Lvov-Volyn Basin – G	---	1.1 – 5.2	5.5 – 14.0
Dneper Basin – B	50.0 – 55.6	---	14.5 – 27.9
Moscow Basin – B			
Humite	13.2 – 39.2	---	11.8 – 50.7
Sapropelitic – B	7.9 – 31.8	---	5.7 – 40.9
Pechora Basin			
Khalmeryusk deposit – K	---	0.5 – 2.0	9.0 – 34.5
Upper-Syryagin deposit – T	---	1.0 – 2.0	8.0 – 40.0
Lower-Syryagin deposit – Zh-G	---	0.7 – 2.0	8.0 – 50.0
Vorkuta Deposit – Zh	---	1.4 – 2.5	10.0 – 44.0
Yunyagin deposit – K	---	0.8 – 1.3	9.0 – 34.0
Vorgashor deposit – G	---	1.0 – 4.7	8.0 – 43.0
Usinsk deposit – Zh	---	1.1 – 2.7	10.0 – 35.5
G	---	1.5 – 2.9	20.0 – 60.0
Intin deposit – D	---	5.0 – 7.2	22.0 – 30.0
Kama Basin – D-G	12.0 – 15.0	2.0 – 5.3	9.6 – 23.9
CAUCASUS			
Tkibuli deposit – Zh-G	---	4.0 – 10.0	23.0 – 40.0
Tkvarchal deposit – Zh-G	---	0.7 – 1.9	11.0 – 30.0
Akhaltsikh deposit – B	-- – 35.0	---	-- – 35.0
Other – B	---	---	13.0 – 45.0
Deposits of the Northern Slope of the Caucasus			
Khumarin deposit – D-G	---	---	12.0 – 19.0
Amgatin and Kubano-Malkin deposits – G-Pzh	---	---	12.0 – 30.0
Malo-Labinsk – Zh	---	---	7.1 – 14.0
Bolshe-Labinsk – K-Zh	---	---	12.4
Tolstobugorsk – K	---	---	5.0 – 30.0
URALS			
Kizel – D-G-Zh	---	2.3 – 5.8	18.0 – 40.0
South Ural Basin – B	50.0 – 60.0	---	10.0 – 42.0
Chelyabinsk Basin – B	11.0 – 15.0	---	17.0 – 32.0
Orsk (East Ural) Basin – B	27.0 – 40.0	---	12.0 – 27.0
North Sosvinsk Basin – B			
Serov deposit – B	---	11.1 – 15.5	17.8 – 30.9
Bogoslov deposit – B	---	-- – 14.0	-- – 20.0

Sulfur Content Absolutely Dry Coal %	Volatiles %	C %	H %	Calorie Content of the Combustible Mass – (Ash and Moisture Free) Cal/kg.
1.4 – 6.0	40 – 46	74.0 – 79.2	5.1 – 5.7	7,400 – 7,900
1.3 – 7.5	35 – 44	78.4 – 82.9	5.0 – 5.8	7,650 – 8,400
1.0 – 6.0	24 – 35	82.4 – 87.0	4.5 – 5.5	8,250 – 8,600
0.9 – 4.9	18 – 26	86.7 – 90.7	4.0 – 5.4	8,450 – 8,750
0.8 – 3.6	13 – 18	87.1 – 91.2	3.9 – 5.1	8,300 – 8,700
0.9 – 6.0	8 – 15	88.0 – 92.4	3.8 – 4.6	8,300 – 8,650
1.0 – 5.0	2 – 7.2	89.4 – 96.4	1.2 – 3.0	7,950 – 8,350
0.7 – 4.0	33 – 39	79.5 – 81.5	4.7 – 5.0	7,735 – 7,870
1.0 – 5.6	40.5 – 60.2	57.3 – 69.0	5.2 – 6.6	4,380 – 6,921
1.0 – 5.0	19 – 46.5	56.8 – 75.7	3.8 – 6.5	5,823 – 7,177
1.0 – 4.5	48.3 – 86.0	69.2 – 79.4	5.5 – 10.2	4,060 – 9,159
0.5 – 3.5	14.0 – 29.0	87.0 – 90.0	4.5 – 5.6	8,600
0.5 – 1.5	14.0 – 18.0	90.5 – 92.0	3.5 – 4.5	---
0.5 – 1.0	25.5 – 41.0	35.0 – 90.0	4.2 – 5.2	8,400 – 8,620
0.4 – 4.5	27.5 – 39.0	79.0 – 87.0	5.0 – 5.2	8,000 – 8,600
0.5 – 5.2	22.0 – 24.0	85.5 – 89.0	4.6 – 5.1	8,630 – 8,770
0.7 – 2.0	33.0 – 42.0	84.0 – 85.5	5.0 – 5.1	---
0.3 – 8.3	30.0 – 36.0	82.4 – 87.0	4.3 – 5.6	8,300 – 8,600
1.0 – 10.0	32.0 – 45.0	82.4 – 85.0	4.3 – 5.6	8,300 – 8,500
1.3 – 3.5	40.0 – 45.0	69.0 – 78.0	5.0	7,500
3.2 – 5.3	31.0 – 50.0	---	---	7,400
0.7 – 5.0	26.0 – 40.0	73.0 – 77.0	6.0 – 7.0	7,000
0.7 – 1.8	33.0 – 39.0	---	---	---
-- – 2.5	12.0 – 20.0	-- – 65.7	-- – 5.05	---
0.5 – 7.0	---	---	---	6,000
0.4 – 0.7	-- – 41.7	---	---	7,600 – 7,700
0.4 – 1.5	---	---	---	5,000 – 9,000
---	20.7 – 28.5	---	---	8,094
---	18.9	---	---	8,100
---	19.5	---	---	8,100
4.0 – 8.0	29.0 – 47.0	---	---	7,600 – 8,190
1.0 – 6.0	50.0 – 75.0	---	---	5,240 – 7,330
0.9 – 1.4	37.0 – 51.0	71.0 – 76.0	4.7 – 5.8	7,150
---	25.0 – 39.0	---	---	5,500 – 6,500
0.4 – 1.0	44.7 – 50.3	56.0 – 73.4	4.5 – 5.9	5,785 – 6,421
-- – 0.5	-- – 48.0	---	---	6,000 – 6,500

TABLE III (Continued)

Basins and Deposits by Type of Coal	Moisture Content of Coal Total %	Dry %	Ash Content Absolutely Dry Coal %
Volchan deposit – B	---	13.0 – 16.0	12.0 – 23.0
Egorshino – T-A	---	0.8 – 3.0	20.5 – 20.8
Bulansh deposit – G	-- – 7.0	-- – 3.2	-- – 18.0
Elkinsk deposit – D	---	-- – 5.0	-- – 19.0
Ural-Caspian Basin – B	---	10.5 – 21.9	17.2 – 35.5
Dombarovski deposit – A	---	2.5 – 5.5	16.0 – 45.0
WESTERN AND EASTERN SIBERIA (Southern Part)			
Gorlov Basin – A	---	-- – 2.1	1.5 – 7.0
Kuznetsk Basin – A	-- – 8.0	---	-- – 11.0
T	-- – 6.0	---	-- – 16.7
K	4.0 – 5.0	---	10.0 – 15.0
SS	4.0 – 8.5	---	9.5 – 15.5
Zh	5.0 – 6.0	---	8.5 – 12.0
G	5.0 – 8.0	---	8.0 – 11.5
D	8.5 – 26.0	---	5.0 – 8.0
Kansk-Achinsk – B	32.0 – 42.0	12.0 – 25.0	7.0 – 20.0
Minusinsk Basin – D-G	-- – 13.0	-- – 3.0	7.0 – 21.0
Ulukhem Basin – G-Zh	---	0.5 – 2.4	6.0 – 25.2
Irkutsk Basin – D-G	-- – 14.0	2.0 – 7.0	10.0 – 20.0
B	25.0 – 28.0	12.0 –	12.0 – 45.0
EASTERN SIBERIA (Northern Part)			
Taimyr Basin			
Slobodsk deposit – T	---	-- – 1.8	17.7 – 18.2
Pyasina – K-T	---	-- – 1.9	8.6 – 12.2
Uglenosnoi River – K-T	---	-- – 3.1	3.9 – 7.4
Tunguska Basin – K-Zh	---	-- – 1.0	10.0 – 34.0
G-D	---	-- – 5.0	6.1 – 17.7
Ust-Yenisey Basin – D	14.7 – 15.6	---	5.3 – 6.7
South-Yakutsk Basin			
Chulmakansk deposit – Zh	---	-- – 1.2	10.4 – 17.2
Neryungrinsk deposit – K	---	-- – 1.0	15.0 – 20.0
Lena Basin			
Anabar-Khatanga – B	---	12.0 – 14.0	9.0 – 20.0
Taimyrlyrsk deposit – D	---	-- – 5.0	-- – 6.0
Chai-Tumus deposit – D-G	---	1.4 – 4.5	5.0 – 38.0
Ogoner-Yuryakh – G-Zh	---	1.2 – 3.3	15.0 – 23.0
Yakutsk region – B	32.0 – 36.0	---	8.0 – 37.0
Ust-Vilyuisk region – B	---	-- – 6.9	-- – 10.7
Nyurbinsk region			
Kempendyaisk deposit – B	---	-- – 11.0	-- – 7.0
Ust-Markhin deposit – B	---	-- – 12.0	-- – 27.0
Kirov deposit – B	---	-- – 10.0	-- – 37.0
Spornoe deposit – B	---	-- – 14.0	-- – 21.0
Verkhoyansk region – D	---	4.0 – 5.0	7.0 – 14.0
G-Zh	---	0.8 – 1.8	9.0 – 12.0
Lower Aldan deposits – B	---	-- – 8.0	-- – 13.0
Tiksi deposit – B	-- – 46.0	-- – 6.0	-- – 6.0

Sulfur Content Absolutely Dry Coal %	Volatiles %	C %	H %	Calorie Content of the Combustible Mass — (Ash and Moisture Free) Cal/kg.
0.3 - 0.7	40.0 - 50.0	---	---	6,300 - 6,600
0.4 - 1.8	3.5 - 13.4	88.3 - 91.7	2.8 - 4.7	6,512 - 8,523
-- - 1.2	34.0 - 49.0	---	---	7,900 - 7,950
-- - 0.8	39.0 - 52.0	-- - 67.7	-- - 5.9	5,500 - 7,800
7.2 - 9.0	24.3 - 38.2	---	---	---
0.1 - 5.7	---	---	---	---
-- - 0.3	1.5 - 3.5	-- - 92.0	-- - 2.2	---
-- - 0.6	-- - 5.0	---	---	8,600
-- - 0.6	9.0 - 11.0	-- - 89.5	-- - 4.2	8,500
0.4 - 0.5	9.0 - 27.0	87.0 - 89.0	4.6 - 5.0	8,400 - 8,600
0.3 - 0.4	23.0 - 32.0	82.0 - 87.0	5.0 - 5.5	8,000 - 8,500
0.6 - 0.8	31.0 - 36.5	84.0 - 86.0	-- - 5.5	8,400 - 8,600
0.3 - 0.7	39.0 - 42.0	82.0 - 83.0	-- - 5.8	8,100 - 8,250
0.3 - 1.2	40.0 - 45.0	-- - 80.0	-- - 6.0	7,000 - 7,850
-- - 0.8	45.0 - 50.0	67.0 - 75.0	1.6 - 9.6	6,500 - 6,800
0.5 - 1.3	35.0 - 46.0	---	---	7,300 - 8,200
0.1 - 0.9	38.8 - 49.1	---	---	8,500 - 8,900
0.5 - 8.0	40.0 - 45.0	78.0 - 81.0	5.0 - 6.8	7,600 - 8,100
1.0 - 2.0	50.0 - 55.0	69.0 - 77.0	6.5 - 7.0	7,000 - 7,200
0.87 - 1.2	8.0 - 12.0	89.8 - 92.8	2.3 - 3.5	8,160 - 8,164
0.4 - 0.9	16.7 - 25.0	87.1 - 88.8	3.6 - 5.0	8,330 - 8,430
0.4 - 0.7	10.0 - 26.0	80.3 - 88.7	3.3 - 5.1	8,000 - 8,007
-- - 1.5	---	---	---	7,000 - 8,400
0.2 - 0.6	33.0 - 39.0	---	---	5,531 - 7,760
0.3 - 0.5	39.0 - 43.1	72.1 - 75.2	4.1 - 4.8	6,774 - 6,829
0.3 - 0.4	31.1 - 35.8	80.0 - 90.0	4.0 - 5.7	---
-- - 1.0	18.0 - 22.0	---	---	8,500
-- - 0.3	-- - 39.5	---	---	5,133 - 5,970
-- - 0.5	36.0 - 43.0	---	---	7,650
---	24.7 - 36.0	---	---	7,472 - 7,787
-- - 0.5	38.0 - 40.0	---	---	7,630 - 8,160
0.3 - 1.6	42.0 - 48.0	70.0 - 75.0	5.0 - 6.4	---
-- - 0.4	-- - 49.5	---	---	6,340
-- - 1.0	-- - 47.9	---	---	6,700
-- - 0.5	39.0 - 52.0	---	---	6,000 - 6,600
0.1 - 4.4	30.0 - 67.0	---	---	---
---	43.0 - 50.0	---	---	---
0.3 - 0.5	47.0 - 50.0	---	---	7,000 - 8,100
0.3 - 0.6	15.0 - 40.0	---	---	8,000 - 8,600
0.2 - 0.5	37.0 - 40.0	---	---	7,300 - 7,400
-- - 0.4	-- - 46.0	---	---	6,700

TABLE III (Continued)

Basins and Deposits by Type of Coal	Moisture Content of Coal Total %	Dry %	Ash Content Absolutely Dry Coal %
THE NORTHEAST			
Zyryansk coal area – K	---	-- – 2.9	-- – 9.0
Zh	---	2.0 – 4.0	1.5 – 23.0
D	---	-- – 6.5	-- – 13.3
Momski deposit – Zh	---	-- – 5.0	-- – 11.0
Arkagalinsk coal area – D-G	-- – 15.0	---	-- – 10.0
Elgen deposit – B	45.0 – 50.0	---	11.0 – 50.0
Bukhta-Ugolnaya – G (?)	6.3 – 8.4	---	-- – 18.0
Avekov deposit – B	---	6.0 – --	6.0 – 49.0
TRANSBAIKAL			
Lake Gusin deposit – B	-- – 21.0	-- – 5.0	15.0 – 20.0
Tarbagataisk deposit – B	---	-- – 8.0	12.0 – 25.0
Chernovsk deposit – B	-- – 33.0	-- – 11.0	10.0 – 19.0
Kharanorsk deposit – B	-- – 33.0	-- – 21.0	8.0 – 14.0
Arbagaro-Kholbonsk – B	-- – 25.0	-- – 11.0	-- – 15.0
Bukachachin deposit – G-D	---	7.0 – 11.0	-- – 12.0
FAR EAST			
Middle Amur region – B	-- – 37.0	---	8.0 – 20.0
Bureya (Bureinsk) – G	---	0.7 – 2.0	19.0 – 48.0
Suifunsk Basin – D	5.0 – 8.5	0.6 – 2.8	12.0 – 27.0
Suchan Basin – T-Ps	6.0 – 7.0	-- – 5.0	21.0 – 27.0
Uglov Basin-Artemov – B	-- – 28.0	-- – 16.0	25.0 – 35.0
Maikhinsk coal area – B	-- – 25.0	---	4.0 – 40.0
Suputinsk deposit – B	---	-- – 16.0	22.0 – 40.0
Bikinsk basin – B	---	-- – 19.0	22.0 – 35.0
SAKHALIN – B	16.0 – 25.0	---	7.0 – 36.0
D	3.5 – 9.5	---	4.0 – 35.0
KAZAKHSTAN			
Karaganda Basin – K-Zh	---	0.8 – 3.5	10.0 – 43.0
B	-- – 30.0	---	10.0 – 25.0
Ubagan Basin – B	30.0 – 35.0	5.8 – 11.0	6.0 – 45.0
Ekibastuz deposit – K-Pzh-G	---	5.0 – 12.0	14.0 – 35.0
Maikuben deposit – B	20.0 – 25.0	---	8.0 – 32.0
Lenger deposit – B	---	---	17.0 – 25.0
MIDDLE ASIA			
Hissar, South Hissar, and Zeravshan – G-Pzh	---	---	2.7 – 30.9
Kugitang – Ps	---	---	0.9 – 27.9
Ziddin, Shuroabad - Ravnous deposits – Pzh	---	---	6.0 – 34.0
South Tadzhik – G-Pzh	---	---	2.0 – 15.2
Mionadus deposit – T-Pzh	---	---	18.2 – 33.1
Sulyukta deposit – B	---	6.0 – 18.0	6.0 – 35.0
Angren deposit – B	-- – 30.0	---	-- – 16.0

Sulfur Content Absolutely Dry Coal %	Volatiles %	C %	H %	Calorie Content of the Combustible Mass – (Ash and Moisture Free) Cal/kg.
-- – 0.4	-- – 27.8	-- – 87.8	-- – 4.8	8,000
0.2 – 0.5	26.0 – 43.0	---	---	8,600
-- – 0.5	38.5 – 45.6	-- – 76.5	-- – 5.4	7,319 – 7,689
-- – 0.5	-- – 27.7	---	---	7,600
-- – 0.4	-- – 38.0	---	---	7,850
---	53.0 – 60.0	---	---	6,200 – 6,450
0.5 – 6.5	40.0 – 50.0	---	---	8,000 – 8,500
0.5 – 1.6	---	-- – 68.6	---	---
---	---	---	---	7,000
2.5 – 4.0	---	---	---	-- – 7,200
-- – 1.0	---	---	---	-- – 7,200
-- – 1.0	---	---	---	-- – 6,200
-- – 2.0	---	---	---	-- – 6,750
-- – 1.0	38.0 – 42.0	---	---	-- – 8,100
-- – 0.03	39.0 – 42.0	70.0 – 71.0	3.5 – 4.5	6,200 – 6,500
-- – 1.0	37.0 – 44.0	77.0 – 82.0	4.2 – 6.0	7,000 – 8,500
-- – 0.4	-- – 50.0	---	---	7,800 – 8,200
-- – 1.0	4.0 – 38.0	80.0 – 83.0	4.0 – 5.0	6,000 – 8,500
-- – 0.6	-- – 48.0	-- – 72.0	-- – 6.0	7,000 – 7,150
0.1 – 0.6	---	---	---	6,200 – 7,200
-- – 1.0	-- – 54.0	-- – 69.0	-- – 5.0	-- – 6,700
-- – 0.6	42.0 – 58.0	---	---	6,000 – 7,000
0.2 – 1.0	42.0 – 52.0	68.0 – 75.0	5.0 – 5.9	6,650 – 7,600
0.2 – 0.9	42.0 – 49.0	75.0 – 81.5	5.6 – 6.5	7,600 – 8,000
0.4 – 3.5	18.0 – 36.0	84.5 – 90.0	4.2 – 5.0	8,370 – 8,870
0.4 – 1.7	-- – 40.0	72.0 – 73.0	4.7 – 5.3	6,200 – 7,100
0.6 – 2.0	35.0 – 50.0	---	---	6,600 – 7,100
0.5 – 2.5	23.0 – 32.0	79.0 – 86.0	4.7 – 5.4	7,500 – 8,000
-- – 1.0	40.0 – 45.0	75.0 – 77.0	---	6,500 – 7,000
2.5 – 5.0	35.0 – 40.0	70.0 – 72.0	3.5 – 4.5	6,600 – 7,000
0.7 – 1.0	27.2 – 32.6	84.1 – 86.8	5.3 – 5.6	8,070 – 8,480
0.5 – 3.0	5.5 – 12.6	90.2 – 95.2	3.5 – 4.9	8,480 – 8,730
---	22.0 – 35.0	79.0 – 81.0	4.7 – 7.4	7,190 – 8,100
0.2 – 0.8	4.7 – 6.8	89.5 – 93.7	2.7 – 3.5	8,140 – 8,600
---	18.1 – 24.8	---	---	-- – 7,575
---	---	---	---	-- – 7,000
---	-- – 33.0	---	---	-- – 7,125

TABLE III (Continued)

Basins and Deposits by Type of Coal	Moisture Content of Coal		Ash Content Absolutely Dry Coal %
	Total %	Dry %	
Shurab deposit – B	---	---	5.0 – 38.0
Minkush coal area – B	---	---	10.7 – 16.8
Kyzyl-Kiya deposit – B	-- – 27.0	---	8.2 – 18.8
Kok-Yangak deposit – D	-- – 15.0	---	-- – 17.7
East Fergana (Uzgen) Basin			
Kumbel deposit – G-D	---	-- – 2.3	9.3 – 19.7
Karagasha deposit – G	---	-- – 1.7	5.0 – 16.0
Tuyuk deposit – G	---	1.2 – 4.0	4.0 – 11.0
Kok-Kiya deposit – Zh	---	-- – 1.3	4.0 – 14.0
Kara-Tyube deposit – T	---	-- – 0.6	8.0 – 12.0
Chitty deposit – (unknown)	---	-- – 0.8	10.2 – 20.2
Besh-Terek deposit – Zh	---	-- – 0.7	10.0 – 20.0
Tarielga deposit – G	---	-- – 2.0	15.0 – 19.0
Zindan deposit – G	---	-- – 2.0	5.0 – 21.0
Fan-Yagnob deposit – G	---	---	3.0 – 15.0
Dzhergalan deposit – D	5.9 – 15.0	2.8 – 31.6	3.5 – 16.0

Sources

N. V. Shabrova and A. V. Tyzhnova, *Zapazy Uglei i Goryuchikh Slantsev*, (Reserves of Coal and Combustible Shale) (Moscow: Gosgeoltekhizdat, 1958), pp. 31–161.

S. V. Troyanski, ed., *Geologiya Ugol'nykh Mestorozhdeniy i Marksheiderskoe Delo* (The

Sulfur Content Absolutely Dry Coal %	Volatiles %	C %	H %	Calorie Content of the Combustible Mass – (Ash and Moisture Free) Cal/kg.
-- – 1.8	36.0 – 42.0	---	---	6,600 – 7,300
1.2 – 3.6	35.3 – 40.3	-- – 73.6	-- – 4.2	5,900 – 7,030
1.0 – 2.0	32.6 – 40.0	---	---	7,000 – 7,200
-- – 1.7	28.7 – 41.0	---	---	7,023 – 7,792
0.3 – 1.0	37.0 – 40.0	78.0 – 81.0	4.7 – 5.5	7,400 – 8,200
0.5 – 0.9	25.0 – 38.0	84.0 – 86.0	5.8 – 6.2	8,270 – 8,480
0.5 – 1.0	24.0 – 32.0	84.0 – 88.0	4.5 – 6.2	7,600 – 8,700
0.6 – 0.9	13.0 – 37.9	86.1 – 86.6	5.8 – 7.0	8,300 – 8,670
0.3 – 0.6	6.3 – 11.0	90.8 – 92.6	4.3 – --	-- – 8,700
-- – 0.7	6.9 – --	92.0 – --	3.3 – --	-- – 8,400
0.5 – 1.3	22.0 – 26.0	85.0 – 87.0	4.8 – 5.0	8,430 – 8,610
2.4 – 3.2	31.0 – 32.0	80.0 – 84.0	---	7,629 – 7,811
0.4 – --	29.0 – 34.0	80.0 – 86.0	4.1 – 5.0	8,194 – 8,230
0.3 – 1.4	21.0 – 40.0	---	---	-- – 8,300
0.2 – 1.8	35.8 – 44.9	69.4 – 81.9	3.6 – 5.2	6,580 – 8,040

Geology of Coal Deposits and the Mining Industry) (Moskva: Ugletekhizdat, 1957), pp. 190–445.

A. A. Gapeev, *Tverdyye Goryuchiye Iskopayemyye* (Solid Mineral Fuels) (Moskva: Gosgeolizdat, 1949), pp. 124–315.

TABLE IV

PRODUCTION OF SOVIET COAL BASINS, UNDER THE MINISTRY OF THE COAL INDUSTRY OF THE U.S.S.R., FOR SELECTED YEARS 1913-1955 BY AMOUNT, TYPE, PERCENT, METHOD OF MINING, AND PRODUCTIVITY PER WORKER(1)

Basins, Combinats, and Trusts Categories	1913	1927/28	1932	1937
MINISTRY OF THE COAL INDUSTRY, U.S.S.R.				
Total production ths/T(2)	29,117.0	33,912.0	62,307.0	124,562.0
in percent	100.0	100.0	100.0	100.0
Coal (Bit.) ths/tons(3)	23,209.0	22,851.0	37,877.0	79,458.2
in percent	79.7	67.4	61.0	63.8
Anthracite ths/tons	4,778.0	8,003.0	18,139.0	26,946.7
in percent	16.4	23.6	29.1	21.6
Brown coal ths/tons	1,130.0	3,058.0	6,191.0	18,157.1
in percent	3.9	9.0	9.9	14.6
Underground ths/tons	28,932.0	33,612.0	61,941.0	122,057.8
in percent	99.4	99.1	99.4	98.0
Open pit ths/tons	185.0	300.0	366.0	2,504.2
in percent	0.6	0.9	0.6	2.0
Tons per worker-month	12.8	12.7	16.2	26.9
MINISTRY OF THE COAL INDUSTRY OF THE UKRAINE				
Total production ths/T	22,760.0	23,138.0	39,269.2	67,458.8
in percent	100.0	100.0	100.0	100.0
Coal (Bit.) ths/tons	22,760.0	17,757.7	25,550.8	47,997.8
in percent	100.0	76.7	65.0	71.1
Anthracite ths/tons	---	5,380.3	13,649.4	19,234.0
in percent	---	23.3	34.8	28.6
Brown coal ths/tons	---	---	69.0	227.0
in percent	---	---	0.2	0.3
Underground ths/tons	22,760.0	23,138.0	39,269.2	67,458.8
in percent	100.0	100.0	100.0	100.0
Open pit ths/tons	---	---	---	---
in percent	---	---	---	---
Tons per worker-month	---	---	---	---
DONETS BASIN				
Total production ths/T	25,288.0	25,732.0	43,847.0	75,041.0
in percent	100.0	100.0	100.0	100.0
Coal (Bit.) ths/tons	20,513.0	17,824.0	25,888.0	48,292.0
in percent	81.1	69.3	59.0	64.4
Anthracite ths/tons	4,775.0	7,908.0	17,959.0	26,749.0
in percent	18.9	30.7	41.0	35.6
Underground ths/tons	25,288.0	25,732.0	43,847.0	75,041.0
in percent	100.0	100.0	100.0	100.0
Tons per worker-month	12.5	12.0	14.8	23.7

(1)Source: *Ugolnaya Promyshlennost S.S.S.R., Statisticheskiy Sprobochnik* (Moskva: Ugeltekhizdat, 1957), 368 pp. Computed from data on pages: 42-52, and 232-233. See also: Dmitriy Grigorevich Onika, *Pod-Moskovnyy Ugolnyy Bassein, (1855-1955)* (Moskva: Moskovskiy Rabochiy, 1956), 235 pp. especially p. 147; and Tsentralnoe Statisticheskoe Upravlenie

Years						
1940	1945	1950	1955	1956	1958 Tons 10^6	1965 (Projected)
153,192.6	142,933.3	248,855.9	376,498.7		495.8	612.0
100.0	100.0	100.0	100.0			
95,329.6	79,300.4	136,320.6	206,516.5			
62.2	55.5	54.7	54.8			
32,296.4	15,995.2	38,971.2	57,533.2			
21.1	11.2	15.7	15.3			
25,566.6	47,637.7	73,564.1	112,449.0			
16.7	33.3	29.6	29.9			
146,883.9	125,152.8	221,714.6	311,564.5			
95.9	87.6	89.1	82.8			
6,308.7	17,780.5	27,141.3	64,934.2			
4.1	12.4	10.9	17.2			
30.6	23.8	30.1	37.8			
76,217.2	29,198.6	73,827.1	120,783.4		164.46	211.0
100.0	100.0	100.0	100.0			
52,985.9	18,965.6	48,681.6	74,949.2			
69.5	68.4	65.9	62.1			
22,867.3	9,194.7	24,205.8	37,139.8			
30.0	31.5	32.8	30.7			
364.0	38.3	939.7	8,694.4			
0.5	0.1	1.3	7.2			
76,217.2	29,198.6	73,728.1	113,652.4			
100.0	100.0	99.9	94.1			
---	---	99.0	7,131.0			
---	---	0.1	5.9			
25.7	14.1	21.8	26.2			
85,508.9	36,933.5	89,678.9	135,334.1	154,100.0	181.7	225.4
100.0	100.0	100.0	100.0			
53,407.4	21,282.7	51,172.9	78,185.8			
62.5	57.6	57.1	57.8			
32,101.5	15,650.8	38,506.0	57,148.3			
37.5	42.4	42.9	42.2			
85,508.9	36,933.5	89,678.9	135,334.1			
100.0	100.0	100.0	100.0			
26.1	14.6	22.7	25.8			

pri Sovete Ministrov SSSR, *Dostizheniya Sovetskoy Vlasti Za Sorok Let V Tsifrakh, Statisticheskiy Sbornik* (Moskva: Gosudarstvennoe Statisticheskoe Izdatelstvo, 1957), p. 81.

(2) ths/T and ths/tons = thousands of metric tons.

(3) Coal (Bit.) = Bituminous coal – from the Russian Kamennyy Ugol-hard coal.

TABLE IV (Continued)

Basins, Combinats, and Trusts				
Categories	**1913**	**1927/28**	**1932**	**1937**
Donbas within the Ukraine				
Total production ths/tons	22,760.0	23,138.0	39,200.2	67,231.8
in percent	100.0	100.0	100.0	100.0
Coal (Bit.) ths/tons	---	17,757.7	25,550.8	47,997.8
in percent	---	76.7	65.2	71.4
Anthracite ths/tons	---	5,380.3	13,649.4	19,234.0
in percent	---	23.3	34.8	28.6
Underground ths/tons	22,760.0	23,138.0	39,200.2	67,231.8
in percent	100.0	100.0	100.0	100.0
Shakhtantratsit				
Total production ths/tons	2,528.0	2,594.0	4,646.8	7,809.2
in percent	100.0	100.0	100.0	100.0
Coal (Bit.) ths/tons	---	66.3	337.2	294.2
in percent	---	2.6	7.3	3.8
Anthracite ths/tons	---	2,527.7	4,309.6	7,515.0
in percent	---	97.4	92.7	96.2
Underground ths/tons	2,528.0	2,594.0	4,646.8	7,809.2
in percent	100.0	100.0	100.0	100.0
MOSCOW BASIN				
Total production brown coal only – ths/tons	300.0	1,135.0	2,613.0	7,507.0
in percent	100.0	100.0	100.0	100.0
Underground ths/tons	300.0	1,135.0	2,613.0	7,507.0
in percent	100.0	100.0	100.0	100.0
Tons per worker-month	13.0	15.1	20.1	28.3
KUZNETSK BASIN				
Total production coal (Bit.) only – ths/tons	774.0	2,618.0	6,780.0	17,340.0
in percent	100.0	100.0	100.0	100.0
Underground ths/tons	774.0	2,618.0	6,780.0	17,340.0
in percent	100.0	100.0	100.0	100.0
Open pit ths/tons	---	---	---	---
in percent	---	---	---	---
Tons per worker-month	---	19.7	24.2	41.2
PECHORA BASIN				
Total production coal (Bit.) only – ths/tons	---	---	4.0	95.0
in percent	---	---	100.0	100.0
Underground ths/tons	---	---	4.0	95.0
in percent	---	---	100.0	100.0
Tons per worker-month	---	---	---	---
BASINS OF THE URALS				
Total production ths/tons	1,217.0	2,084.0	3,166.0	8,085.0
in percent	100.0	100.0	100.0	100.0
Coal (Bit.) ths/tons	897.0	1,109.0	1,529.0	3,712.2
in percent	73.8	55.7	48.3	46.0

Years						
1940	1945	1950	1955	1956	1958 Tons 10^6	1965 (Projected)
75,853.2	29,160.3	72,887.4	111,883.9		149.3	189.5
100.0	100.0	100.0	100.0			
52,985.9	19,965.6	48,681.6	74,744.9			
69.9	68.5	66.8	66.8			
22,867.3	9,194.7	24,205.8	37,139.1			
30.1	31.5	33.2	33.2			
75,853.2	29,160.3	72,887.4	111,883.9			
100.0	100.0	100.0	100.0			
9,655.7	7,773.2	16,791.5	23,450.2		32.4	35.9
100.0	100.0	100.0	100.0			
421.5	1,317.1	2,491.3	3,441.7			
4.4	16.9	14.8	14.7			
9,234.2	6,456.1	14,300.2	20,008.5			
95.6	83.1	85.2	85.3			
9,655.7	7,773.2	16,791.5	23,450.2			
100.0	100.0	100.0	100.0			
9,948.8	20,021.4	30,621.6	39,301.5	42,200.0	47.2	35.8
100.0	100.0	100.0	100.0			
9,948.8	20,021.4	30,621.6	39,301.5			
100.0	100.0	100.0	100.0			
32.4	23.4	35.2	42.0			
21,137.1	28,994.0	36.814.1	56,537.1	66,200.0	75.3	103.0
100.0	100.0	100.0	100.0	100.0		
21,137.0	28,994.0	36,069.7	50,872.0			
100.0	100.0	98.0	90.0			
---	---	744.4	5,665.1			
---	---	2.0	10.0			
43.1	34.0	36.3	46.7			
261.8	3,318.7	8,687.4	14,153.4	15,400.0	16.8	19.0
100.0	100.0	100.0	100.0	100.0		
261.8	3,318.7	8,687.4	14,153.4			
100.0	100.0	100.0	100.0			
---	17.3	25.1	37.7			
11,700.9	25,066.7	32,174.8	46,856.6	52,300.0	61.04	58.8
100.0	100.0	100.0	100.0	100.0		
4,567.4	7,720.8	10,372.3	11,676.4			
39.0	30.8	32.2	24.9			

TABLE IV (Continued)

Basins, Combinats, and Trusts Categories	1913	1927/28	1932	1937
Anthracite ths/tons	3.0	95.5	180.0	197.7
in percent	0.2	4.8	5.7	2.4
Brown coal ths/tons	317.0	785.0	1,457.0	4,175.1
in percent	26.0	39.5	46.0	51.6
Underground ths/tons	1,032.0	1,689.0	2,800.0	6,700.8
in percent	84.8	84.9	88.4	82.9
Open pit ths/tons	185.0	300.0	366.0	1,384.2
in percent	15.2	15.1	11.6	17.1
Tons per worker-month	---	18.7	16.2	31.0
Molotovugol				
Total production coal (Bit.) underground only – ths/tons	897.0	1,109.0	1,529.0	3,712.2
in percent	100.0	100.0	100.0	100.0
Chelyabinskugol				
Total production brown coal only – ths/tons	132.0	485.0	1,091.0	3,464.3
in percent	100.0	100.0	100.0	100.0
Underground ths/tons	132.0	485.0	1,091.0	2,789.8
in percent	100.0	100.0	100.0	80.5
Open pit ths/tons	---	---	---	674.5
in percent	---	---	---	19.5
Sverdlovskugol				
Total production ths/tons	188.0	395.0	546.0	907.4
in percent	100.0	100.0	100.0	100.0
Coal (Bit.) ths/tons	---	---	---	---
in percent	---	---	---	---
Anthracite ths/tons	3.0	95.0	180.0	197.7
in percent	1.6	24.1	33.0	21.8
Brown coal ths/tons	185.0	300.0	366.0	709.7
in percent	98.4	75.9	67.0	78.2
Underground ths/tons	3.0	95.0	180.0	197.7
in percent	1.6	24.1	33.0	21.8
Open pit ths/tons	185.0	300.0	366.0	709.7
Bashkirugol				
Total production brown coal only – ths/tons Open pit only	---	---	---	---
KARAGANDA BASIN (Karagandaugol)				
Total production ths/tons	---	---	722.0	3,937.0
in percent	---	---	100.0	100.0
Coal (Bit.) ths/tons	---	---	722.0	3,860.0
in percent	---	---	100.0	98.0
Brown coal ths/tons	---	---	---	77.0
in percent	---	---	---	2.0

*Including the output of coal in the Ekibastuz deposits. These deposits also in the Kazakh Republic are under the Karagandaugol Combinat.

Years						
1940	1945	1950	1955	1956	1958 Tons 10^6	1965 (Projected)
194.9	344.4	465.2	384.9			
1.7	1.4	1.5	0.8			
6,938.6	17,001.5	21,337.3	34,795.3			
59.3	67.8	66.3	74.3			
8,273.4	12,459.0	18,096.6	23,195.1			
70.7	49.7	56.2	49.5			
3,427.5	12,607.7	14,078.2	23,661.5			
29.3	50.3	43.8	50.5			
43.9	40.3	43.9	61.1			
4,530.2	7,667.7	10,127.6	11,020.9			
100.0	100.0	100.0	100.0			
5,540.4	11,266.4	12,274.0	17,639.6			
100.0	100.0	100.0	100.0			
3,492.9	4,393.8	7,259.1	11,133.8			
63.0	39.0	59.1	63.1			
2,047.5	6,872.6	5,014.9	6,505.8			
37.0	61.0	40.9	36.9			
1,574.9	6,132.6	9,773.2	16,361.4			
100.0	100.0	100.0	100.0			
---	53.1	244.7	655.5			
---	0.9	2.5	4.0			
194.9	344.4	465.2	384.9			
12.4	5.6	4.8	2.4			
1,380.0	5,735.1	9,063.3	15,321.0			
87.6	93.5	92.7	93.6			
194.9	397.5	709.9	1,040.4			
12.4	6.5	7.3	6.4			
1,380.0	5,735.1	9,063.3	15,321.0			
---	---	---	1,834.7			
6,297.9	11,265.5	16,304.6	26,812.1	25,900.0	30.4†	48.6†
100.0	100.0	100.0	100.0	100.0		
6,180.9	9,412.0	12,087.9	20,750.5*			
98.1	83.5	74.1	77.4			
117.0	1,853.5	4,216.7	6,061.6			
1.9	16.5	25.9	22.6			

†Includes 6.1×10^6 tons for Ekibastuz in 1958 and a projected 10×10^6 tons in 1965.

TABLE IV (Continued)

Basins, Combinats, and Trusts				
Categories	**1913**	**1927/28**	**1932**	**1937**
Underground ths/tons	---	---	722.2	3,937.0
in percent	---	---	100.0	100.0
Open pit ths/tons	---	---	---	---
in percent	---	---	---	---
Tons per worker-month	---	---	11.7	30.9
DEPOSITS OF MIDDLE ASIA				
Total production ths/tons	248.0	271.0	720.0	1,044.0
in percent	100.0	100.0	100.0	100.0
Coal (Bit.) ths/tons	121.0	20.0	196.0	239.0
in percent	48.8	7.4	27.2	22.9
Brown coal ths/tons	127.0	251.0	524.0	805.0
in percent	51.2	92.6	72.8	77.1
Underground ths/tons	248.0	271.0	720.0	1,044.0
in percent	100.0	100.0	100.0	100.0
Open pit ths/tons	---	---	---	---
in percent	---	---	---	---
Tons per worker-month	---	10.6	12.0	16.5
DEPOSITS OF EASTERN SIBERIA (Vostsibugol)				
Total production ths/tons	847.0	1,009.0	2,448.0	5,730.0
in percent	100.0	100.0	100.0	100.0
Coal (Bit.) ths/tons	566.0	720.0	1,902.0	4,088.0
in percent	66.8	71.4	77.7	71.3
Brown coal ths/tons	281.0	289.0	546.0	1,642.0
in percent	33.2	28.6	22.3	28.7
Underground ths/tons	847.0	1,009.0	2,448.0	5,730.0
in percent	100.0	100.0	100.0	100.0
Open pit ths/tons	---	---	---	---
in percent	---	---	---	---
Tons per worker-month	---	31.6	27.2	37.2
DEPOSITS OF THE FAR EAST				
Total production ths/tons	373.0	1,073.0	1,702.0	4,700.0
in percent	100.0	100.0	100.0	100.0
Coal (Bit.) ths/tons	268.0	475.0	725.0	976.0
in percent	71.8	44.3	42.6	20.8
Brown coal ths/tons	105.0	598.0	977.0	3,724.0
in percent	28.2	55.7	57.4	79.2
Underground ths/tons	373.0	1,073.0	1,702.0	3,580.0
in percent	100.0	100.0	100.0	76.2
Open pit ths/tons	---	---	---	1,120.0
in percent	---	---	---	23.8
Tons per worker-month	---	15.0	33.6	31.2
Primorskugol				
Total production ths/tons	340.0	1,063.0	1,426.0	2,941.0
in percent	100.0	100.0	100.0	100.0

* Including the output of coal in the Ekibastuz deposits. These deposits also in the Kazakh Republic are under the Karagandaugol Combinat.

Years						
1940	1945	1950	1955	1956	1958 Tons 10^6	1965 (Projected)
6,297.9	9,412.0	12,087.9	18,468.1			
100.0	83.5	74.1	68.9			
---	1,853.5	4,216.7	8,344.0*			
---	16.5	25.9	31.1*			
43.9	29.7	40.4	62.4			
1,919.7	1,689.1	4,237.0	6,333.1			
100.0	100.0	100.0	100.0			
421.9	367.3	701.4	984.0			
22.0	21.7	16.6	15.5			
1,497.8	1,321.8	3,535.6	5,349.1			
78.0	78.3	83.4	84.5			
1,919.7	1,689.1	3,161.2	4,209.8			
100.0	100.0	74.6	66.5			
---	---	1,075.8	2,123.3			
---	---	25.4	33.5			
22.2	13.0	24.0	34.8			
8,527.5	7,644.0	15,062.4	23,173.3		35.96	56.0
100.0	100.0	100.0	100.0			
6,773.4	5,585.1	11,001.2	16,214.4			
79.4	73.1	73.0	70.0			
1,754.1	2,058.9	4,061.2	6,958.9			
20.6	26.9	27.0	30.0			
8,009.8	7,049.1	12,814.9	12,290.0			
93.9	92.2	85.1	53.0			
517.7	594.9	2,247.5	10,883.3			
6.1	7.8	14.9	47.0			
51.6	38.8	47.6	77.0			
6,594.8	7,019.7	12,047.5	16,057.1		20.0	25.8
100.0	100.0	100.0	100.0			
1,663.6	1,962.8	3,582.0	4,915.8			
25.2	28.0	29.7	30.6			
4,931.2	5,056.9	8,465.5	11,141.3			
74.8	72.0	70.3	69.4			
4,231.3	4,295.3	7,367.8	8,931.1			
64.2	61.2	61.2	55.6			
2,363.5	2,724.4	4,679.7	7,126.0			
35.8	38.8	38.8	44.4			
34.6	35.7	40.2	48.4			
3,327.1	3,362.2	4,857.7	5,441.8			
100.0	100.0	100.0	100.0			

TABLE IV (Continued)

Basins, Combinats, and Trusts				
Categories	**1913**	**1927/28**	**1932**	**1937**
Coal (Bit.) ths/tons	235.0	465.0	449.0	657.0
in percent	69.1	43.7	31.5	22.3
Brown coal ths/tons	105.0	598.0	977.0	2,284.0
in percent	30.9	56.3	68.5	77.7
Underground ths/tons	340.0	1,063.0	1,426.0	2,941.0
in percent	100.0	100.0	100.0	100.0
<u>Dalvostugol</u>				
Total production ths/tons	---	---	---	1,440.0
in percent	---	---	---	100.0
Coal (Bit.) ths/tons	---	---	---	---
in percent	---	---	---	---
Brown coal ths/tons	---	---	---	1,440.0
in percent	---	---	---	100.0
Underground ths/tons	---	---	---	320.0
in percent	---	---	---	22.2
Open pit ths/tons	---	---	---	1,120.0
in percent	---	---	---	77.8
<u>Sakhalinugol</u>				
Total production ths/tons	33.0	10.0	276.0	319.0
in percent	100.0	100.0	100.0	100.0
Coal (Bit.) ths/tons	33.0	10.0	276.0	319.0
in percent	100.0	100.0	100.0	100.0
Brown coal ths/tons	---	---	---	---
in percent	---	---	---	---
DEPOSITS OF THE GEORGIAN S.S.R. (Gruzugol)				
Total production ths/tons	70.0	85.0	205.0	398.0
in percent	100.0	100.0	100.0	100.0
Coal (Bit.) ths/tons	70.0	85.0	205.0	398.0
in percent	100.0	100.0	100.0	100.0
Brown coal ths/tons	---	---	---	---
in percent	---	---	---	---
Underground ths/tons	70.0	85.0	205.0	398.0
in percent	100.0	100.0	100.0	100.0
Tons per worker-month	---	6.7	16.5	18.5

Years						
1940	1945	1950	1955	1956	1958 Tons 10^6	1965 (Projected)
1,172.2	1,351.1	2,033.5	2,225.7			
35.2	40.2	41.9	40.9			
2,154.9	2,011.1	2,824.2	3,216.1			
64.8	59.8	58.1	59.1			
3,337.1	3,362.2	4,857.7	5,441.8			
100.0	100.0	100.0	100.0			
2,782.2	3,045.8	4,936.9	6,992.4			
100.0	100.0	100.0	100.0			
5.9	---	94.4	457.6			
0.2	---	1.9	6.5			
2,776.3	3,045.8	4,842.5	6,534.8			
99.8	100.0	98.1	93.5			
418.7	321.4	581.9	457.6			
15.0	10.6	11.8	6.5			
2,363.5	2,724.4	4,355.0	6,534.8			
85.0	89.4	88.2	93.5			
485.5	611.7	2,252.9	3,622.9			
100.0	100.0	100.0	100.0			
485.5	611.7	1,454.1	2,232.5			
100.0	100.0	64.5	61.6			
---	---	798.8	1,390.4			
---	---	35.5	38.4			
618.0	650.1	1,717.8	2,706.2		3.01	3.8
100.0	100.0	100.0	100.0			
618.0	650.1	1,671.8	2,559.3			
100.0	100.0	97.3	94.6			
---	---	46.0	146.9			
---	---	2.7	5.4			
618.0	650.1	1,717.8	2,706.2			
100.0	100.0	100.0	100.0			
22.4	16.6	24.2	33.4			

TABLE V

SHIPMENTS OF COAL FROM THE DONETS, KUZNETSK, KARAGANDA, AND KIZEL BASINS TO THE CONSUMING REGIONS, IN TONS AND PER CENT OF THE TOTAL PRODUCTION OF THE BASIN[a]

Consuming Region	1940	1950	1955
Donets Basin			
The South (Incl. Crimea)	51,279,687 59.97%	52,515,926 58.56%	82,716,201 61.47%
Incl. Stalinsk and Voroshilovgrad Oblasts	25,001,353 29.25%	25,540,550 28.48%	42,697,908 31.55%
Central Region	18,213,395 21.30%	18,536,628 20.67%	30,057,703 22.21%
Northern Caucasus	4,429,361 5.18%	7,317,798 8.16%	9,608,721 7.10%
Incl. Rostov and Kamensk Oblasts	2,513,063 2.96%	4,053,486 4.52%	5,995,300 4.43%
Volga Region	2,975,709 3.48%	7,308,830 8.15%	4,614,892 3.41%
The West	1,333,938 1.56%	1,112,018 1.24%	4,208,890 3.14%
The Northwest	6,575,634 7.69%	1,909,160 2.13%	2,706,682 2.00%
Trans-Caucasus	76,958 0.09%	914,724 1.02%	1,353,341 1.00%
North	607,113 0.71%	17,935 0.02%	54,133 0.04%
Other Regions	17,101 0.02%	44,839 0.05%	13,533 0.01%
Kuznetsk Basin			
Western Siberia	10,987,064 51.98%	17,759,121 48.24%	23,717,313 41.95%
Urals	5,871,886 27.78%	14,504,755 39.40%	21,382,331 37.82%
Middle Asia and Kazakhstan	2,508,973 11.87%	3,806,557 10.34%	3,126,501 5.53%
Incl. the Region South of Alma-Ata	902,554 4.27%	1,553,555 4.22%	435,335 0.77%
Central Region	574,929 2.27%	184,070 0.50%	4,415,547 7.81%
Volga Region	1,166,767 5.52%	559,574 1.52%	3,810,600 6.74%
Other Regions	27,478 0.13%	--- ---	84,805 0.15%

TABLE V (Continued)

Consuming Region	1940	1950	1955
Karaganda Basin			
Urals	3,458,806 54.92%	8,147,408 49.97%	11,926,022 44.48%
Kazakhstan and Middle Asia	2,052,485 32.59%	6,101,181 37.42%	10,754,333 40.11%
Incl. the Region South of Alma-Ata	--- ---	358,701 0.46%	329,788 7.06%
Central Region	306,077 4.86%	53,805 0.33%	943,785 3.52%
Volga Region	480,529 7.63%	1,643,503 10.08%	2,858,169 10.66%
Western Siberia	--- ---	358,701 2.20%	329,788 1.23%
Kizel Basin			
Urals	3,980,686 87.87%	8,171,960 80.69%	8,245,837 74.82%
Central Region	53,999 11.92%	1,785,495 17.63%	2,413,577 21.90%
Volga Region	--- ---	140,773 1.93%	52,900 0.48%
Other Regions	9,513 0.21%	29,370 0.29%	308,585 2.80%

[a]Computed from data on p. 40 of D. T. Onika, *Ugolnaya Promyshlennost S.S.S.R. v Shestoi Pyatileke* (Moskva: Ugletekhizdat, 1956), 62 pp., and data in Coal Appendix Table IV.

OIL SHALE APPENDIX

TABLE I

ENERGY POTENTIAL OF THE MINEABLE RESERVES OF OIL SHALE IN THE U.S.S.R., FOR JANUARY 1, 1955

Basins, Deposits and Regions	Mineable Reserves 10^6[a]	Calorie Content Mineable Reserves 10^{12}[b]	Kwt. Hour Content Mineable Reserves 10^6[c]
TOTAL U.S.S.R.	54,574.8	103,289	120,573,197
BALTIC BASIN	14,220	27,018	31,416,530
Estonia	10,463	19,879	23,116,115
Gdov	3,551	6,746	7,845,294
Chudov	206	391	455,119
EAST RUSSIAN PLATFORM	16,794	31,906	37,103,320
Ivhem	6,350	12,065	14,029,182
Sysolsk	5,401	10,261	11,932,537
Kostroma	42	79	92,791
Tatar-ulyanovsk	1,029	1,955	2,272,390
Kuibyshev-Saratov	3,972	7,546	8,775,419
KAZAKHSTAN	704.8	1,338	1,557,128
Kenderal	698	1,326	1,542,105
Baikhazhin	6.8	13	15,023
KUZNETSK BASIN			
Barzasski	500	950	1,104,660
NORTHEAST SIBERIA	22,356	42,476	49,391,558

[a]Computed from data in N. V. Shabarova and A. V. Tyzhnova, *Zapasy Uglei I Goryuchikh Slantsev S.S.S.R.* (Moskva: Gosgeoltekhizdat, 1958), pp. 162-176.

[b]Computed on a basis of 1900 kilogram calories as outlined in V. Kalamkarov, "Osnovye Napravleniya v Razbitin Proizodstva Popliva v Shestom Pyapiletii," *Planovoe Khozyaistvo*, 1957, p. 17.

[c]Conversion factor 0.0011628 from N. A. Lange, *Handbook of Chemistry*, *op. cit.*, p. 1721 and data in column 3 above.

www.ingramcontent.com/pod-product-compliance
Ingram Content Group UK Ltd.
Pitfield, Milton Keynes, MK11 3LW, UK
UKHW022127220426
470267UK00006B/25

9 780837 184913